Essential Statistics

Essential Statistics

Second edition

D.G. Rees

Oxford Polytechnic
Department of Computing and Mathematical Sciences

CHAPMAN & HALL
London · New York · Tokyo · Melbourne · Madras

**Published by Chapman & Hall, 2-6 Boundary Row,
London SE1 8HN, UK**

Chapman & Hall, 2-6 Boundary Row, London SE1 8HN, UK

Blackie Academic & Professional, Wester Cleddens Road, Bishopbriggs, Glasgow G64 2NZ, UK

Chapman & Hall Inc., One Penn Plaza, 41st Floor, New York, NY 10119, USA

Chapman & Hall Japan, Thomson Publishing Japan, Hirakawacho Nemoto Building, 6F, 1-7-11 Hirakawa-cho, Chiyoda-ku, Tokyo 102, Japan

Chapman & Hall Australia, Thomas Nelson Australia, 102 Dodds Street, South Melbourne, Victoria 3205, Australia

Chapman & Hall India, R. Seshadri, 32 Second Main Road, CIT East, Madras 600 035, India

First edition 1985
Second edition 1989
Reprinted 1990, 1991 (twice), 1992, 1994

Typeset in 10/12 Times by Best-set Typesetter Ltd, Hong Kong
Printed in Great Britain by T.J. Press (Padstow) Ltd, Padstow, Cornwall

ISBN 0 412 32030 4

A catalogue record for this book is available from the British Library
Library of Congress Cataloging-in-Publication Data
Rees, D.G.
Essential statistics/D.G. Rees.-2nd ed.
p. cm.
Bibliography:p.
Includes index.
ISBN 0412 32030 4
1. Statistics. I. Title
QA276.12R44 1989 89-31953
519.5-dc20 CIP

Contents

Preface to Second Edition xi

Preface to First Edition xii

Acknowledgements xiv

1 What is statistics? 1

1.1 Statistics as a science 1
1.2 Types of statistical data 2
Worksheet 1 4

2 Some statistical notation 6

2.1 Σ 6
2.2 Factorials 7
2.3 x^y 8
2.4 e^x 8
2.5 Decimal places and significant figures 8
Worksheet 2 9

3 Summarizing data by tables and by graphical methods 11

3.1 Tables for one continuous variable 11
3.2 Table for one discrete variable 13
3.3 Table for one categorical variable 14
3.4 When to summarize one-variable data in a table 14
3.5 Tabular methods for two-variable data 14
3.6 Graphical methods for one continuous variable 16
3.7 Graphical method for one discrete variable 17
3.8 Graphical method for one categorical variable 18
3.9 Graphical methods for two-variable data 19
3.10 Stem and leaf displays 20
3.11 Summary 21
Worksheet 3 22

4 Summarizing data by numerical measures 24

4.1 Averages 24
4.2 Sample mean ($\bar{x}$) 25
4.3 Sample median 26
4.4 Sample mode 27
4.5 When to use the mean, median and mode 28
4.6 Measures of variation 30
4.7 Sample standard deviation (s) 30
4.8 Sample inter-quartile range 33
4.9 When to use standard deviation and inter-quartile range 33
4.10 Other measures of variation 34
4.11 Box and whisker plots 34
4.12 Measures of shape 35
4.13 Summary 35
Worksheet 4 36

5 Probability 39

5.1 Introduction 39
5.2 Basic ideas of probability 40
5.3 The *a priori* definition of probability for equally likely outcomes 40
5.4 The relative frequency definition of probability, based on experimental data 41
5.5 The range of possible values for a probability value 42
5.6 Probability, percentage, proportion and odds 42
5.7 Subjective probability 43
5.8 Probabilities involving more than one event 43
5.9 Multiplication law (the 'and' law) 43
5.10 Addition law (the 'or' law) 45
5.11 Mutually exclusive and exhaustive events 46
5.12 Complementary events and the calculation of P(at least 1 . . .) 47
5.13 Probability trees 47
5.14 Summary 49
Worksheet 5 49

6 Discrete probability distributions 52

6.1 Introduction 52
6.2 Binomial distribution, an example 52
6.3 The general binomial distribution 53
6.4 Calculating binomial probabilities, an example 54
6.5 The mean and standard deviation of the binomial distribution 55

6.6 Using tables to obtain binomial probabilities 56
6.7 Poisson distribution, an introduction 57
6.8 Some examples of Poisson variables 57
6.9 The general Poisson distribution 58
6.10 Calculating Poisson probabilities, an example 58
6.11 The mean and standard deviation of the Poisson distribution 60
6.12 Using tables to obtain Poisson probabilities 60
*6.13 Poisson approximation to the binomial distribution 60
6.14 Summary 61
Worksheet 6 61

7 Continuous probability distributions 65

7.1 Introduction 65
7.2 The normal distribution 67
7.3 An example of a normal distribution 68
7.4 Comparison of different normal distributions 70
7.5 Rectangular distribution 71
*7.6 The normal approximation to the binomial distribution 72
7.7 Summary 73
Worksheet 7 73

8 Samples and populations 77

8.1 Introduction 77
8.2 Reasons for sampling 77
8.3 Sampling methods 78
8.4 Sample size 80
8.5 Sampling distribution of the sample mean 81
8.6 Summary 82
Worksheet 8 83

9 Confidence interval estimation 85

9.1 Introduction 85
9.2 95% confidence intervals 85
9.3 Calculating a 95% confidence interval for the mean, μ, of a population – large sample size, n 86
9.4 Calculating a 95% confidence interval for the mean, μ, of a population – small sample size, n 88
*9.5 The t-distribution 89
9.6 The choice of sample size when estimating the mean of a population 90
*9.7 Degrees of freedom 90
9.8 95% confidence interval for a binomial probability 91

9.9 The choice of sample size when estimating a binomial probability 92
9.10 95% confidence interval for the mean of a population of differences, 'paired' samples data 93
9.11 95% confidence interval for the difference in the means of two populations, 'unpaired' samples data 94
9.12 Summary 95
Worksheet 9 96

10 Hypothesis testing 100
10.1 Introduction 100
10.2 What is a hypothesis? 101
10.3 Which is the null hypothesis and which is the alternative hypothesis? 101
10.4 What is a significance level? 102
10.5 What is a test statistic, and how do we calculate it? 102
10.6 How do we find the tabulated test statistic? 103
10.7 How do we compare the calculated and tabulated test statistics? 103
10.8 What is our conclusion, and what assumptions have we made? 104
10.9 Hypothesis test for the mean, μ, of a population 104
10.10 Two examples of hypothesis tests with one-sided alternative hypotheses 105
10.11 Hypothesis test for a binomial probability 106
10.12 Hypothesis test for the mean of a population of differences, 'paired' samples data 108
10.13 Hypothesis test for the difference in the means of two populations, 'unpaired' samples data 109
*10.14 The effect of choosing significance levels other than 5% 110
10.15 What if the assumptions of a hypothesis test are not valid? 110
10.16 The connection between confidence interval estimation and hypothesis testing 110
10.17 Summary 111
Worksheet 10 112

11 Non-parametric hypothesis tests 116
11.1 Introduction 116
11.2 Sign test for the median of a population 117
11.3 Sign test for the median of a population of differences, 'paired' samples data 118
*11.4 Sign test for large samples ($n > 10$) 119

11.5 Wilcoxon signed rank test for the median of a population of differences, 'paired' samples data 120
*11.6 Wilcoxon signed rank test for large samples ($n > 25$) 122
11.7 Mann–Whitney U test for the difference in the medians of two populations, 'unpaired' samples data 123
*11.8 Mann–Whitney U test for large samples ($n_1 > 20, n_2 > 20$) 125
11.9 Summary 126
Worksheet 11 126

12 Association of categorical variables 130

12.1 Introduction 130
12.2 Contingency tables 130
12.3 χ^2 test for independence, 2×2 contingency table data 131
12.4 Further notes on the χ^2 test for independence, contingency table data 133
12.5 χ^2 test for independence, 3×3 table 134
12.6 Summary 136
Worksheet 12 136

13 Correlation of quantitative variables 139

13.1 Introduction 139
13.2 Pearson's correlation coefficient 140
13.3 Hypothesis test for Pearson's population correlation coefficient, ϱ 142
13.4 The interpretation of significant and non-significant correlation coefficients 143
13.5 Spearman's rank correlation coefficient 146
13.6 Hypothesis test for Spearman's rank correlation coefficient 148
13.7 Summary 148
Worksheet 13 149

14 Regression analysis 153

14.1 Introduction 153
14.2 Determining the regression equation from sample data 154
14.3 Plotting the regression line on the scatter diagram 155
14.4 Predicting values of y 156
*14.5 Confidence intervals for predicted values of y 156
*14.6 Hypothesis test for the slope of the regression line 159
*14.7 The connection between regression and correlation 160
*14.8 Transformations to produce linear relationships 161
14.9 Summary 162
Worksheet 14 162

15 χ^2 goodness-of-fit tests 167

15.1 Introduction 167
15.2 Goodness-of-fit for a 'simple proportion' distribution 167
*15.3 Goodness-of-fit for a binomial distribution 169
*15.4 Goodness-of-fit for a Poisson distribution 171
*15.5 Goodness-of-fit for a normal distribution 173
15.6 Summary 176
Worksheet 15 176

16 Minitab 181

16.1 Introduction 181
16.2 Getting started 181
16.3 Data input 181
16.4 Editing data 182
16.5 Summarizing data by graphical methods and numerical summaries 183
16.6 Saving data for a future Minitab session, and leaving Minitab 184
16.7 The HELP command 185
16.8 Getting hard-copy printouts 185
16.9 Discrete probability distributions 186
16.10 Continuous probability distributions 188
16.11 Simulation of the sampling distribution of the sample mean 189
16.12 Confidence interval estimation for means 190
16.13 Simulation and confidence intervals 191
16.14 Hypothesis testing for means 192
16.15 Non-parametric hypothesis tests 194
16.16 χ^2 test for independence, contingency table data 194
16.17 Scatter diagrams and correlation 195
16.18 Regression analysis 196
16.19 χ^2 goodness-of-fit tests 197
16.20 Summary 198
Worksheet 16 200

Appendix A Multiple choice test 202
Appendix B Solutions to worksheets and multiple choice test 207
Appendix C Glossary of symbols 230
Appendix D Statistical tables 233
Appendix E Further reading 252

Index 253

Preface to the Second Edition

The main feature of this new edition is a substantial addition (Chapter 16) on applications of the interactive statistical computer package, Minitab. This package has become widely used in colleges as an aid to teaching statistics. The new chapter contains over 20 sample programs illustrating how Minitab can be used to draw graphs, calculate statistics, carry out tests and perform simulations. The chapter could act as a primer for first-time Minitab users.

There are also new sections in Chapters 3 and 4 on some aspects of exploratory data analysis. Some changes have been made to the statistical tables. For example, Tables D.1 and D.2 now give cumulative probabilities in terms of 'r or fewer ...' instead of 'r or more ...'. The tables are now consistent with those adopted by most GCSE examination boards and also with the output from the Minitab **CDF** command for both the binomial and Poisson distributions. For similar reasons Table D.3(a) now gives the cumulative distribution function for the normal distribution, i.e. areas to the left of various values of z. Another change is that the conditions for the use of the normal approximation to the binomial have been brought into line with accepted practice. There are other minor changes too numerous to list here.

I am grateful for the opportunity to update and enhance the successful first edition. Many thanks to all those who have expressed their appreciation of *Essential Statistics* as a course text or who have made helpful suggestions for improvements.

D.G. Rees

Preface to the First Edition

TO THE STUDENT

Are you a student who requires a basic statistics text-book? Are you studying statistics as part of a study of another subject, for example one of the natural, applied or social sciences, or a vocational subject? Do you have an O-level or GCSE in mathematics or an equivalent qualification? If you can answer 'yes' to all three questions I have written this book primarily for you.

The main aim of this book is to encourage and develop your interest in statistics, which I have found to be a fascinating subject for over twenty years. Other aims are to help you to:

1. Understand the essential ideas and concepts of statistics.
2. Perform some of the most useful statistical methods.
3. Be able to judge which method is the most appropriate in a given situation.
4. Be aware of the assumptions and pitfalls of the methods.

Because of the wide variety of subject areas which require knowledge of introductory statistics, the worked examples of the various methods given in the main part of the text are not aimed at any one subject. In fact they deliberately relate to methods which can be applied to 'people data' so that every student can follow them without specialist knowledge. The end-of-chapter Worksheets, on the other hand, relate to a wide variety of subjects to enable different students to see the relevance of the various methods to their areas of special interest.

You should tackle each worksheet before proceeding to the next chapter. To help with the necessary calculations you should be, or quickly become, familiar with an electronic hand calculator with the facilities given below.† (These facilities are now available on most scientific calculators.)

† *Calculators* The minimum requirements are: a memory, eight figures on the display, a good range of function keys (including square, square root, logarithm, exponential, powers, factorials) and internal programs for mean and standard deviation.

Answers and partial solutions are given to all the questions on the worksheets. When you have completed the whole book (except for the sections marked with an asterisk (*), which may be omitted at the first reading), a multiple choice test is also provided, as a quick method of self-assessment.

TO THE TEACHER OR LECTURER

This book is not intended to do away with face-to-face teaching of statistics. Although my experience is that statistics is best taught in a one-to-one situation with teacher and student, this is clearly not practical in schools, colleges and polytechnics where introductory courses in statistics to non-specialist students often demand classes and lectures to large groups of students. Inevitably these lectures tend to be impersonal.

Because I have concentrated on the essential concepts and methods, the teacher who uses this book as a course text is free to emphasize what he or she considers to be the most important aspects of each topic, and also to add breadth or depth to meet the requirements of the particular course being taught.

Another advantage for the teacher is that, since partial solutions are provided to all the questions on the worksheets, students can attempt these questions with relatively little supervision.

WHAT THIS BOOK IS ABOUT

After introducing 'Statistics as a science' in Chapter 1 and statistical notation in Chapter 2, Chapters 3 and 4 deal with descriptive or summary statistics, while Chapters 5, 6 and 7 concentrate on probability and four of the most useful probability distributions.

The rest of the book comes broadly under the heading of statistical inference. After discussing sampling in Chapter 8, two branches of inference – confidence interval estimation and hypothesis testing – are introduced in Chapters 9 and 10 by reference to several 'parametric' cases. Three non-parametric hypothesis tests are discussed in Chapter 11.

In Chapters 12 and 13 association and correlation for bivariate data are covered. Simple linear regression is dealt with in Chapter 14 and χ^2 goodness-of-fit tests in Chapter 15.

I have attempted throughout to cover the concepts, assumptions and pitfalls of the methods, and to present them clearly and logically with the minimum of mathematical theory.

Acknowledgements

The quotations given at the beginning of Chapters 1, 2, 3, 4, 8, 10, 11 and 13 are taken from a very interesting book on diseases and mortality in London in the 18th century. I would like to thank Gregg International, Amersham, England for permission to use these quotations from *An Arithmetical and Medical Analysis of the Diseases and Mortality of the Human Species* by W. Black (1973).

Acknowledgements for the use of various statistical tables are given in Appendix D.

Thanks also to all the colleagues and students who have influenced me, and have therefore contributed indirectly to this book. Most of all I am grateful to my wife, Merilyn, not only for producing the typescript but also for her support and encouragement throughout.

What is statistics?

> Authors. . . have obscured their works in a cloud of figures and calculation: the reader must have no small portion of phlegm and resolution to follow them throughout with attention: they often tax the memory and patience with a numerical superfluity, even to a nuisance.

1.1 STATISTICS AS A SCIENCE

You may feel that the title of this chapter should be, 'What are statistics?' indicating the usual meaning of statistics as numerical facts or numbers. So, for example, the unemployment statistics are published monthly giving the number of people who have received unemployment benefit during the month. However, in the title of this chapter the singular noun 'statistics' is used to mean the science of collecting and analysing data where the plural noun 'data' means numerical or non-numerical facts or information.

We may collect data about 'individuals', that is individual people or objects. There may be many characteristics which vary from one individual to another. We call these characteristics variables. For example, individual people vary in height and employment status, and so height and employment status are variables.

Let us consider an example of some data which we might wish to collect and analyse. Suppose our variable of interest is the height of students at the start of their first year in higher education. We would expect these heights to vary. We could start by choosing one college from all the colleges of higher education, we could choose 50 first-year students from that college's enrolment list, and we could measure the heights of these students and calculate the average height. There are many other ways of collecting and analysing such data. Indeed this book is about how surveys like this should be conducted and clearly they cannot be discussed now in detail. But it is instructive to ask some of the questions which need to be considered before such a survey is carried out.

The most important question is, 'What is the purpose of the survey?' The answer to this question will help us to answer other questions – is it better to choose all 50 students from one college or a number from each of a number of colleges? How many students should be selected altogether, and how many from each of the chosen colleges? How do we select a given number of students from a college's enrolment list? What do we do if a selected student refuses to co-operate in the survey? How do we allow for known or suspected differences between male and female student heights? How accurately should the heights be measured? Does the average height of the students selected for the survey tell us all we need to know about their heights? How can we relate the average height of the selected students to the average height of all students at the start of their first year in higher education?

1.2 TYPES OF STATISTICAL DATA

Before we look at how data may be collected and analysed we will consider the different types of statistical data we may need to study. As stated in the Preface, the main part of this book will be concerned with 'people data', so the following list gives some of the variables which may be collected from people:

Sex
Age
Height
Income
Number of brothers and sisters
Religion
Employment status
Birth order

Some of these variables are *categorical*, that is the 'value' taken by the variable is a non-numerical category or class. An example of a categorical variable is religion, with categories Hindu, Catholic, Moslem, and so on. Some variables are quantifiable, that is they can take numerical values. These numerical variables can be further classified as being either continuous, discrete or ranked using the following definitions:

A *continuous variable* can take any value in a given range.
A *discrete variable* can take only certain distinct values in a given range.
A *ranked variable* is a categorical variable in which the categories imply some order or relative position.

Example

Height is an example of a continuous variable since an individual adult human being may have a height anywhere in the range 100 to 200 centimetres, say. We can usually decide that a variable is continuous if it is *measured* in some units.

Example

Number of brothers and sisters (siblings) is an example of a discrete variable, since an individual human being may have 0, 1, 2,. . . , siblings, but cannot have 1.43, for example. We can usually decide that a variable is discrete if it is *counted.*

Example

Birth order is an example of a ranked variable, since an individual human being may be the first-born, second-born, etc., into a family, with corresponding birth order of 1, 2, etc.

Table 1.1 shows the results of applying similar ideas to all the variables in the above list.

Table 1.1 Examples of types of statistical data

Name of variable	*Type of variable*	*Likely range of values or list of categories*
Sex	Categorical	Male, female
Age	Continuous	0 to 100 years
Height	Continuous	100 to 200 cm
Income	Continuous	£20 to £1000 per week
Number of siblings	Discrete	0, 1, 2,. . . , 10
Religion	Categorical	Hindu, Catholic, Moslem, etc.
Employment status	Categorical	Full-time employed, part-time employed, unemployed
Birth order	Ranked	1, 2,. . . , 10

You may feel that the distinction between the continuous and the discrete variable is, in practice, not as clear cut as stated above. For example, most people give their age as a whole number of years so that age appears to be a discrete variable which increases by one at each birthday. The practice of giving one's age approximately for whatever reason, does not alter the fact that age is fundamentally a continuous variable.

Now try Worksheet 1.

WORKSHEET 1: TYPES OF STATISTICAL DATA

For the following decide whether the variable is continuous, discrete, ranked or categorical. Give a range of likely values or a list of categories. The value or category of the variable varies from one 'individual' to another. The individual may or may not be human, as in the first question where the individual is 'light-bulb'. Name the individual in each case.

1. The number of hours of operation of a number of 100 W light-bulbs.
2. The number of current-account balances checked by a firm of auditors each year.
3. The present cost of bed and breakfast in three-star London hotels.
4. The number of rooms-with-bathroom in three-star London hotels.
5. The type of occupation of adult males.
6. The number of failures per 100 hours of operation of a large computer system.
7. The number of hours lost per 100 hours due to failures of a large computer system.
8. The number of cars made by a car company each month.
9. The position of the British entry in the annual Eurovision song contest each year.
10. The annual rainfall in English counties in 1983.
11. The number of earthquakes per year in a European country in the period 1900–1983.
12. The outputs of North Sea oil rigs in 1983.
13. The α-particle count from a radioactive source in 10-second periods.
14. The number of times rats turn right in ten encounters with a T-junction in a maze.
15. The grades obtained by candidates taking A-level mathematics.
16. The colour of people's hair.
17. The presence or absence of a plant species in each square metre of a meadow.

18. The reaction time of rats to a stimulus.
19. The number of errors per page of a balance sheet.
20. The yield of tomatoes per plant in a greenhouse.
21. The constituents found in core samples when drilling for oil.
22. The percentage hydrogen content of gases collected from samples near to a volcanic eruption.
23. The political party people vote for in an election.

Some statistical notation

I have corrected several errors of
preceding calculators...

Note It is not necessary for you to master all the statistical notation in this chapter before you proceed to Chapter 3. However, references to this notation will be made in later chapters within the context of particular statistical methods. Worksheet 2 which follows this chapter is intended to help you to use your calculator and become familiar with the notation.

2.1 Σ

Suppose we have carried out a survey by measuring the heights of a sample of 50 students and we wish to analyse these data. One obvious thing to do would be to calculate the mean height or, to be more precise, the 'sample mean height'. We would simply add up the heights and divide by 50. In order to generalize this idea so that it could be applied to any set of data consisting of a sample of values of one variable for a number of 'individuals', the following statistical notation is found to be useful.

Suppose we have a sample of n individuals (n can be any positive whole number), and let $x_1, x_2, \ldots, x_n$ represent the values of the variable x for the n individuals. Then,

$$\text{sample mean of } x = \frac{x_1 + x_2 + \cdots + x_n}{n}.$$

If, instead of $x_1 + x_2 + \cdots + x_n$, we write

$$\sum_{i=1}^{i=n} x_i$$

we can see that the symbol Σ (the upper-case version of the Greek letter sigma) stands for the operation of summing. In future, then, it will help you to remember what Σ means in statistics if you remember 'capital sigma means sum'. If it is clear how many values are to be summed we simply write:

$$\Sigma x \qquad \text{instead of} \qquad \sum_{i=1}^{i=n} x_i.$$

So Σx means 'sum the x values'. Also, we will use the symbol $\bar{x}$ to denote the 'sample mean of x'. So the formula above may simply be written:

$$\bar{x} = \frac{\Sigma x}{n}.$$

The symbol $\bar{x}$ and other such symbols can usually be found on scientific hand calculators. Both the notation and the calculator are essential tools in the study of statistics.

Here is some other useful notation:

Σx^2 means square each value of x, and then sum.

$(\Sigma x)^2$ means sum the values of x, and then square the total.

$\Sigma(x - \bar{x})$ means subtract $\bar{x}$ from each value of x, and then sum.

Example

The heights in centimetres of a sample of five people are:

150, 200, 180, 160, 170.

Here $n = 5$, and we will use the formula for $\bar{x}$ by replacing x_1 by 150, etc. So

$$\bar{x} = \frac{\Sigma x}{n} = \frac{150 + 200 + 180 + 160 + 170}{5} = \frac{860}{5} = 172 \text{ cm}.$$

In words, the sample mean height is 172 cm.

$$\Sigma x^2 = 150^2 + 200^2 + 180^2 + 160^2 + 170^2 = 149\,400 \text{ cm}^2$$
$$(\Sigma x)^2 = 860^2 = 739\,600 \text{ cm}^2$$
$$\begin{aligned}\Sigma(x-\bar{x}) &= (150-172) + (200-172) + (180-172) + (160-172) + (170-172)\\ &= -22 + 28 + 8 + -12 + -2\\ &= 0.\end{aligned}$$

Note
For any set of values, $\Sigma(x - \bar{x})$ will always be zero.

2.2 FACTORIALS

If n is a positive whole number, the product $1 \times 2 \times 3 \times \cdots \times n$ is called 'factorial n' and is given the symbol $n!$. So

$$n! = 1 \times 2 \times 3 \times \cdots \times n$$

Examples

(a) $3! = 1 \times 2 \times 3 = 6$.
(b) $5! = 1 \times 2 \times 3 \times 4 \times 5 = 120$.
(c) $10! = 1 \times 2 \times 3 \times 4 \times 5 \times 6 \times 7 \times 8 \times 9 \times 10 = 3\,628\,800$.

Many hand calculators have the facility for calculating factorials. Look for a button marked $n!$ or x!.

Zero is the only other number, apart from the positive whole numbers, for which there is a definition of factorial:

$$0! = 1,$$

but $(-5)!$, for example, is not defined and so is meaningless.

Factorials will be used in Chapter 6 and subsequently. For the moment their use is to help you to get to know your calculator.

2.3 x^y

You may already be familiar with squaring a number and using the x^2 facility on your calculator. In Chapter 6 we need to raise numbers to powers other than 2, and this is where the x^y facility is useful.

Example

0.5^4 means use the x^y facility with $x = 0.5$, $y = 4$.

Answer: $0.5^4 = 0.0625$.

2.4 e^x

The letter e on your calculator simply stands for the number 2.718... We will not be interested in the wider mathematical significance of e. It is important in this book only because we need to be able to calculate probabilities in Chapter 6 using the e^x facility on our calculators.

Example

e^{-4} means using the e^x facility for $x = -4$.

Answer: $e^{-4} = 0.0183$, to 4 decimal places.

2.5 DECIMAL PLACES AND SIGNIFICANT FIGURES

Calculators produce many figures on the display and it is tempting to write them all down. You will learn by experience how many figures are meaningful in an answer. For the moment, concentrate on giving answers to a stated number of decimal places or significant figures.

Use the idea that, for example, 3 decimal places (dps) means write three figures only to the right of the decimal point, rounding the third figure (after the decimal point) up if the fourth figure is 5 or more.

Examples

(a) 1.6666 to 3 dps is 1.667.
(b) 1.6665 to 3 dps is 1.667.
(c) 1.6663 to 3 dps is 1.666.
(d) 1.67 to 3 dps is 1.670.
(e) 167 to 3 dps is 167.000.

The number of significant figures means the number of figures (as you scan from left to right) starting with the first non-zero figure. Round the last significant figure up if the figure immediately to its right is 5 or more. Non-significant figures to the left of the decimal point are written as zeros, and to the right of the decimal point are omitted.

Examples

(a) 26 243 to 3 sig. figs. is 26 200.
(b) 2624 to 3 sig. figs. is 2620.
(c) 2626 to 3 sig. figs. is 2630.
(d) 26.24 to 3 sig. figs. is 26.2.
(e) 0.2624 to 3 sig. figs. is 0.262.
(f) 0.002626 to 3 sig. figs. is 0.00263.

WORKSHEET 2: SOME STATISTICAL NOTATION

1. Check that you are able to work out each of these on your calculator.
   (a) $1.3 + 2.6 - 5.7$
   (b) $10.0 - 3.4 - 2.6 - 1.0$
   (c) $(2.3)(14.6)$
   (d) $(0.009)(0.0273)(1.36)$
   (e) $2.3/14.6$
   (f) $1/0.00293$
   (g) $(2.3 + 4.6 + 9.2 + 17.3)/4$
   (h) $28^{0.5}$
   (i) $(0.5)^3$
   (j) $(0.2)^2(0.8)^4$
   (k) $(0.5)^0$
   (l) $(0.2)^{-3}$
   (m) $e^{1.6}$
   (n) $e^{-1.6}$

(o) $13/\sqrt{(10 \times 24)}$
(p) $6 - (-0.5)(4)$
(q) $4!$, $1!$, $6!$, $0!$, $(-3)!$, $(2.4)!$

2. Express the answer to Question:
   (a) 1(c) to 1 dp.
   (b) 1(d) to 2 sig. figs.
   (c) 1(e) to 2 sig. figs.
   (d) 1(f) to 4 sig. figs.
   (e) 1(f) to 1 sig. fig.

3. Use the memory facility on your calculator to work out the following:
   (a) $1 + 2 + 3 + 4 + 5 + 6 + 7 + 8 + 9 + 10$
   (b) $(1 + 2 + 3 + 4 + 5)/5$
   (c) $1^2 + 2^2 + 3^2 + 4^2 + 5^2$
   (d) $(1 \times 2) + (3 \times 4) + (5 \times 6)$.

4. For the eight values of x: 2, 3, 5, 1, 4, 3, 2, 4, find Σx, $\bar{x}$, $(\Sigma x)^2$, Σx^2, $\Sigma(x - \bar{x})$, $\Sigma(x - \bar{x})^2$ and $\Sigma x^2 - \dfrac{(\Sigma x)^2}{n}$.

5. Repeat Question 4 for the five values of x: 2.3, 4.6, 1.3, 7.2, 2.3.

Summarizing data by tables and by graphical methods

> The important data . . . are condensed, classed, and arranged into concise tables.

In this chapter we will consider simple ways of representing data in tables and graphs when the data consist of values either of one or two variables for a number of individuals, which we will refer to as one-variable data or two-variable data.

3.1 TABLES FOR ONE CONTINUOUS VARIABLE

Example

Suppose we have the heights in centimetres (measured to the nearest 0.1 cm) of 50 students at the start of their first year in higher education. We could write down the 50 values as in Table 3.1, but we would find it difficult to make sense of these 'raw' data by looking at this table. It is more informative to form a *grouped frequency distribution* as in Table 3.2.

Table 3.1 A list of the heights (cm) of 50 students

164.8	182.7	168.4	165.6	155.0	176.2	171.2	181.3	176.0	178.4
169.0	161.4	176.8	166.1	159.9	184.0	171.8	169.2	163.7	160.2
173.7	153.6	151.7	165.1	174.2	174.8	173.0	157.8	169.7	179.7
172.2	174.6	179.6	146.3	169.3	164.5	175.0	193.9	184.6	163.6
177.7	190.1	168.7	170.0	165.3	180.1	168.3	185.4	171.3	186.8

Table 3.2 Grouped frequency distribution for heights of 50 students

Height (cm)	Tally marks	Number of students (frequency)
145.0–154.9	|||	3
155.0–164.9	~~||||~~ ||||	9
165.0–174.9	~~||||~~ ~~||||~~ ~~||||~~ ~~||||~~ |	21
175.0–184.9	~~||||~~ ~~||||~~ |||	13
185.0–194.9	||||	4
		Total 50

How is a table such as Table 3.2 formed from the 50 ungrouped values in Table 3.1? First we find the smallest and largest values; for our data these are 146.3 and 193.9. Since this is a range of about 50 cm, a table with five or six groups could be formed, each group covering a range of 10 cm. The lower end-point of the first group, 145, is chosen as a round number just below the smallest value, and this choice determines the lower end-points of the other groups by adding 10 cm each time. The upper end-points of the groups, 154.9, 164.9 and so on, are chosen to leave no gaps and to ensure no overlapping between groups. The final group must contain the largest value, 193.9.

The raw data in Table 3.1 are then read, one at a time, and a tally mark is made in the appropriate group for each height value. When this is completed the tally marks are added to give the number of students or 'frequency' for each group.

Table 3.2 is not the only grouped frequency distribution table which could be drawn for these data. But in drawing up such a table the following guidelines should be followed:

1. Each value must fit into one but only one of the groups. If groups of 145–155, 155–165, etc., had been used in the example, it would not have been clear into which group values of 155, 165, etc., should fit.
2. The number of groups should be between 5 and 15. If there are too few groups the variability in the data (i.e. the way it is distributed) cannot be seen in sufficient detail. If there are too many groups the table becomes less of a summary. In general the larger the number of data values, the larger the number of groups.
3. Equal-width groups are preferable, unless there is some special reason for having groups of unequal width. For the example, the width of each group is 10 cm (= 155 − 145). One advantage of equal-width groups is that it is easier to represent the data graphically (see Section 3.6).

Another table which can be formed for one continuous variable is the *cumulative frequency distribution* table, which can be drawn up from the grouped frequency distribution table (see Table 3.2). In this new table (Table 3.3) we provide information on the number of values less than certain specific values, namely the lower end-points of the groups.

Example

(using the data in Table 3.2)

Table 3.3 Cumulative frequency distribution table for heights of 50 students

Height (cm)	*Cumulative number of students (cumulative frequency)*
145	0
155	3
165	12
175	33
185	46
195	50

3.2 TABLE FOR ONE DISCRETE VARIABLE

If a variable is discrete it may not be desirable to combine values to form groups if the number of values taken by the variable is less than 15.

Example

If the students whose heights were used in the previous section were asked to state their number of siblings, and the only values which resulted were 0, 1, 2, 3 and 4, then Table 3.4 might be the result.

Table 3.4 Grouped frequency distribution table for siblings of 50 students

Number of siblings	*Number of students (frequency)*
0	15
1	20
2	10
3	3
4	2

Note that the guidelines 1 and 2 of the previous section have still been followed. If the number of values taken by the discrete variable is 15 or more it may be sensible to form a smaller number of groups.

3.3 TABLE FOR ONE CATEGORICAL VARIABLE

Example

Suppose we note the hair colour of the 50 students used in previous examples, restricting categories to four, namely black, brown, fair and red. Table 3.5 might be the result.

Table 3.5 Grouped frequency distribution table for hair colour of 50 students

Hair colour	*Number of students (frequency)*
Black	28
Brown	12
Fair	6
Red	4

3.4 WHEN TO SUMMARIZE ONE-VARIABLE DATA IN A TABLE

This is a matter of judgement, but as a rough guide it is perhaps unnecessary to put fewer than 15 values into a grouped frequency table, but necessary with 30 or more values. It depends on whether we feel the table gives a more useful first indication of how the data vary. One point to note is that some information is lost when we group continuous variable data, since we record only how many values occur in each group, not whereabouts each value occurs in its group.

Other methods of summarizing data are by graphical methods and by numerical measures. These methods (which are discussed later in this chapter and in the next) are often used in conjunction with or in preference to tabular methods.

3.5 TABULAR METHODS FOR TWO-VARIABLE DATA

These will not be discussed here in detail but a few specific examples will be given because they will be referred to in later chapters:

(a) When two categorical variables are measured for a number of individuals we may display the data in a two-way or *contingency* table (see Table 3.6). An analysis of this type of data is discussed in Chapter 12.

Table 3.6 Hair colour and temper of 100 individuals

	Temper	
Hair colour	*Bad*	*Good*
Red	30	10
Not red	20	40

(b) When two continuous variables are measured for a relatively small number of individuals we may display the data in two columns or two rows (see Table 3.7). Analysis of this type of data is discussed in Chapters 13 and 14.

Table 3.7 Heights and weights of six individuals

Height (cm)	*Weight* (kg)
170	57
175	64
176	70
178	76
183	71
185	82

(c) When one continuous variable is measured for one individual and the other variable is time, we may display the data in two columns or two rows (see Table 3.8).

Table 3.8 Temperature of an individual at different times

Time	*Temperature* (°F)
8.00	103.0
8.15	102.0
8.30	101.0
8.45	100.0
9.00	99.0
9.15	98.6

(d) When one discrete and one categorical variable are measured for a relatively small number of individuals we may display the data in a number of columns or rows (see Table 3.9). Analysis of this type of 'unpaired' data is discussed in Chapters 9, 10 and 11.

Table 3.9 Examination marks and sex of 14 children

Female marks	85	73	44	30	72	42	36
Male marks	82	59	42	25	54	40	32

3.6 GRAPHICAL METHODS FOR ONE CONTINUOUS VARIABLE

We may represent data from a grouped frequency distribution table for a continuous variable graphically in a *histogram*, as in Fig. 3.1, and the data from a cumulative frequency table in a *cumulative frequency polygon*, as in Fig. 3.2 (sometimes called an *ogive*).

If the groups are of equal width, the vertical axis of a histogram is frequency.† The variable (height, in the example in Fig. 3.1) is represented on the horizontal axis. For each group a rectangle is drawn. The histogram should be a continuous picture; there should be no gaps unless particular groups contain no values. (Strictly speaking the bases of the rectangles should be marked 144.95, 154.95, etc., but we will not worry about such small discrepancies here.) If we look at the shape of the histogram, we see almost the same shape by turning Table 3.2 anticlockwise through 90°. The histogram is simply another way of representing the tally marks.

The vertical axis of a cumulative frequency polygon is cumulative frequency, the horizontal axis represents the variable. Each row of Table 3.3 is represented by a point and the points are joined by straight lines. It is preferable to start with a point at a cumulative frequency of zero. The final point must correspond to the total frequency.

NB Remember to use the group lower end-points to plot the cumulative frequency polygon, not the group mid-points.

The method of drawing the cumulative frequency polygon (unlike the histogram) is the same whether or not the groups are of equal width.

† This is not true if not all the groups are of equal width, since the main property of the histogram is that the area of each rectangle is proportional to the corresponding frequency. If in the above example we had decided to make the last group 185–205 instead of 185–195, that is twice as wide, we should have to halve the height of the rectangle to keep its area the same. The vertical axis in this case should be re-labelled 'Number of students (frequency) per 10 cm group width'.

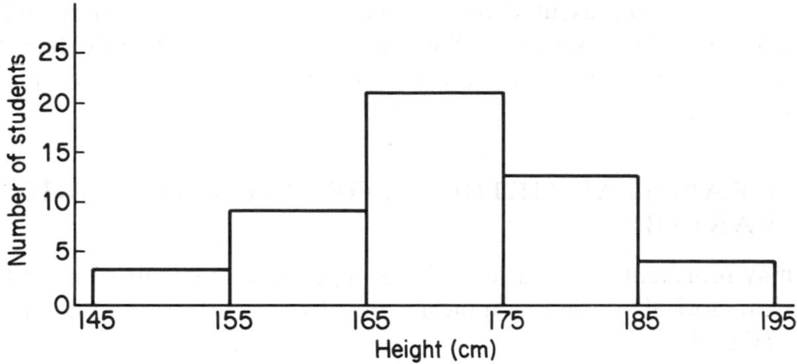

Fig. 3.1 Histogram for the data in Table 3.2.

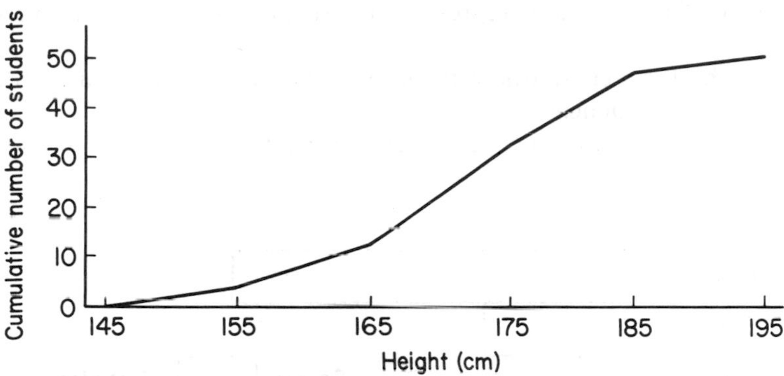

Fig. 3.2 Cumulative frequency polygon for the data in Table 3.3.

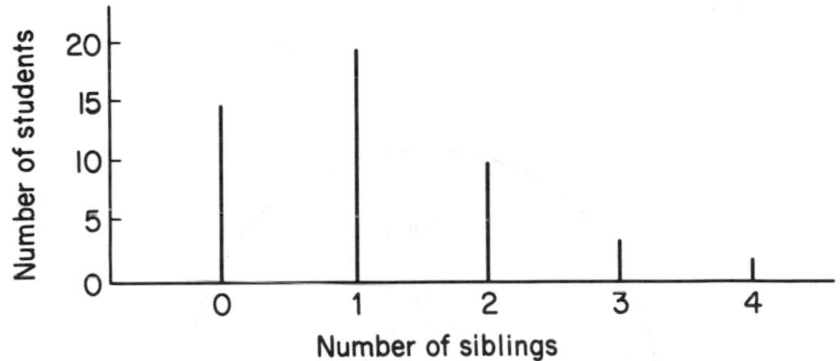

Fig. 3.3 Line chart for the data in Table 3.4.

3.7 GRAPHICAL METHOD FOR ONE DISCRETE VARIABLE

We may represent the data from the grouped frequency distribution table for a discrete variable graphically in a *line chart* (see Fig. 3.3).

Frequency is represented on the vertical axis of a line chart and the variable on the horizontal axis. Since the variable is discrete the line chart is discrete, in the sense that these are gaps between the possible values of the variable.

3.8 GRAPHICAL METHOD FOR ONE CATEGORICAL VARIABLE

We may represent the data from the grouped frequency distribution table for a categorical variable graphically in a *bar chart* (Fig. 3.4) or in a *pie chart* (Fig. 3.5).

To avoid confusion with the histogram, the frequencies in a bar chart are represented horizontally, the length of each bar being proportional to the frequency. The variable is represented vertically, but there is no vertical scale.

A pie chart is a circle where the sizes of the 'pieces of pie' are proportional to the frequencies.

For the example, the angle for 'black' is (28/50) × 360° = 202° and so on.

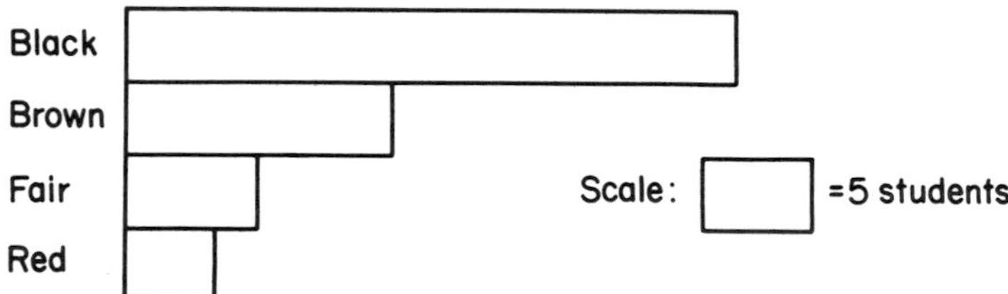

Fig. 3.4 Bar chart for the data in Table 3.5.

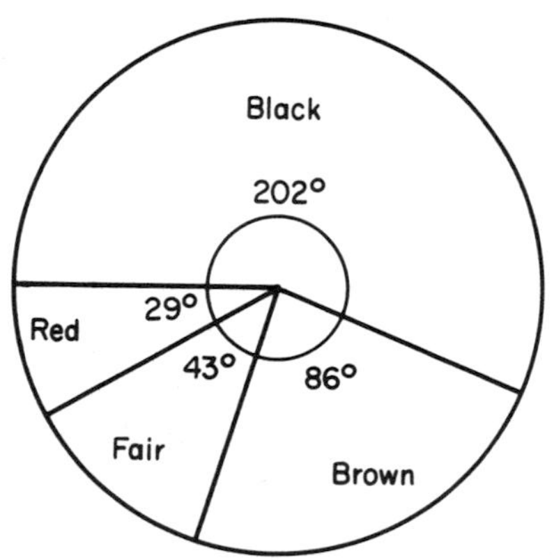

Fig. 3.5 Pie chart for the data in Table 3.5.

3.9 GRAPHICAL METHODS FOR TWO-VARIABLE DATA

We may represent the two-variable data examples of Section 3.5 graphically as in Figs. 3.6, 3.7, 3.8 and 3.9.

Hair colour	*Temper*	
	Bad	*Good*
Red		
Not red		

= 10 individuals

Fig. 3.6 Pictogram for the data in Table 3.6.

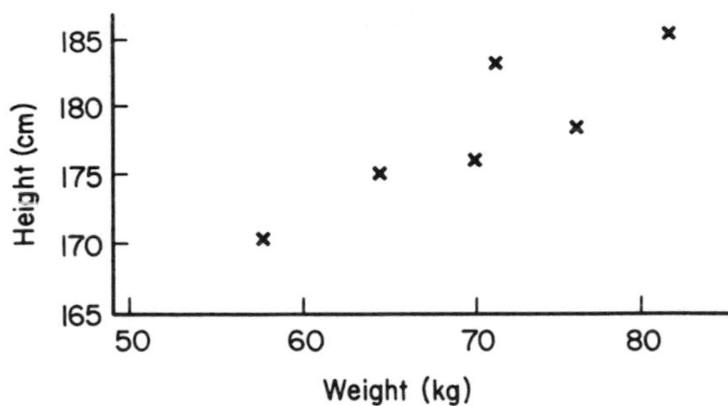

Fig. 3.7 Scatter diagram for the data in Table 3.7.

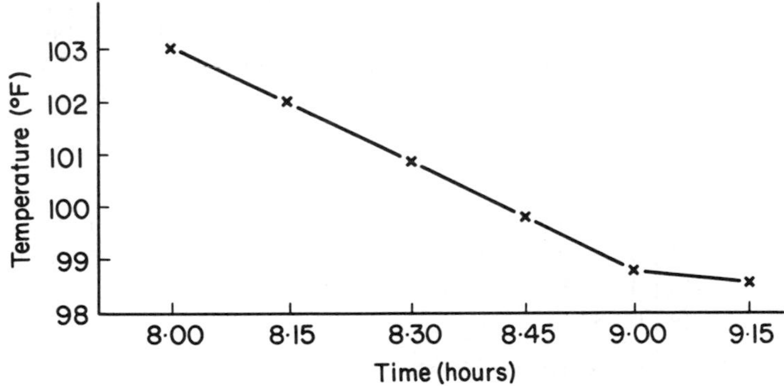

Fig. 3.8 Time-series graph for the data in Table 3.8.

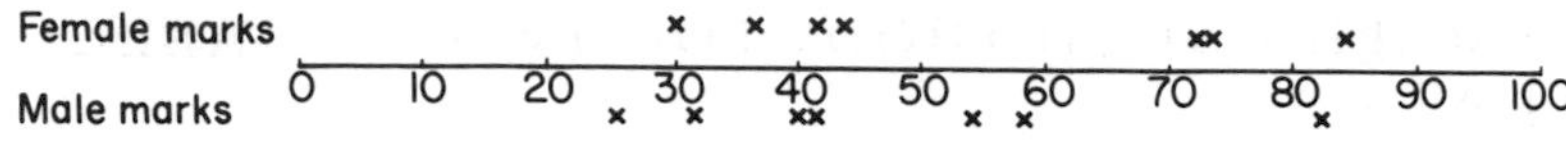

Fig. 3.9 Cross-diagram for the data in Table 3.9.

In a *scatter diagram* one variable is plotted on each axis. Generally, the choice is arbitrary. Each point represents the data for an individual.

In a *time-series graph* the convention is for the vertical scale to represent the variable, and for the horizontal scale to represent time. Since all the data refer to the same individual the points may be joined together.

Where there is insufficient data to plot a separate histogram for the two sets of data (female and male data in the example) a *cross-diagram* is useful for a quick visual comparison between the sets.

3.10 STEM AND LEAF DISPLAYS

A stem and leaf display is a combination of a table and a graph, and is one of a number of techniques which come under the heading of exploratory data analysis, developed by John Tukey (*Exploratory Data Analysis*, Addison-Wesley, 1977). Stem and leaf displays were intended to illustrate continuous variable data, and visually resemble the histogram.

Example

We reproduce the data from Table 3.1 as Table 3.10, but with the height values reduced to three significant figures. A stem and leaf display of these data is shown in Fig. 3.10.

Table 3.10 Heights of 50 students (nearest cm)

165	183	168	166	155	176	171	181	176	178
169	161	177	166	160	184	172	169	164	160
174	154	152	165	174	175	173	158	170	180
172	175	180	146	169	164	175	194	185	164
178	190	169	170	165	180	168	185	171	187

```
        14 | 6
        15 | 2  4  5  8
        16 | 0  0  1  4  4  4  5  5  5  6  6  8  8  9  9  9  9
        17 | 0  0  1  1  2  2  3  4  4  5  5  5  6  6  7  8  8
        18 | 0  0  0  1  3  4  5  5  7
        19 | 0  4
n = 50  14 | 6  represents 146 cm
```

Fig. 3.10 Stem and leaf display for the heights of 50 students.

The column of numbers to the left of the vertical line is called the 'stem', while each digit to the right of the vertical line is a 'leaf'. The different rows of the stem, 14, 15,. . . , 19, are called levels. Notice that the leaves at each level are arranged in rank order from lowest to highest. If you turn Fig. 3.10 anti-clockwise through 90° you will see a shape similar to that of the histogram for these data (Fig. 3.1). One advantage that a stem and leaf display has over the histogram is that it is possible to retrieve the original 'raw' data.

3.11 SUMMARY

When one- or two-variable data are collected for a number of individuals, these data may be summarized in tables or graphically. Some form of grouping is advisable unless there are only a few values. The particular type of table and graph used to summarize the data depends on the type(s) of variable(s). Examples discussed in this chapter are shown in Table 3.11.

Table 3.11

One-variable data

Variable type	*Type of table (and reference)*	*Graph (and reference)*
Continuous	Grouped frequency (Table 3.2) or cumulative frequency (Table 3.3)	Histogram (Fig. 3.1) or cumulative frequency polygon (Fig. 3.2) Stem and leaf displays (Fig. 3.10)
Discrete	Grouped frequency (Table 3.4)	Line chart (Fig. 3.3)
Categorical	Grouped frequency (Table 3.5)	Bar chart (Fig. 3.4) or pie chart (Fig. 3.5)

Two-variable data

Variable type	*Type of table (and reference)*	*Graph (and reference)*
Both categorical	Contingency (Table 3.6)	Pictogram (Fig. 3.6)
Both continuous	Two columns (or rows) (Table 3.7)	Scatter diagram (Fig. 3.7)
One continuous, plus time	Two columns (or rows) (Table 3.8)	Time series (Fig. 3.8)
One discrete, one categorical	Two columns (or rows) (Table 3.9)	Cross-diagram (Fig. 3.9)

WORKSHEET 3: SUMMARIZING DATA BY TABLES AND BY GRAPHICAL METHODS

1. Decide which type of table and graphical method you would use on the following one-variable data sets:
   (a) The number of hours of operation of 100 light-bulbs.
   (b) The type of occupation of 50 adult males.
   (c) The total number of earthquakes recorded in this century for each of ten European countries.
   (d) The percentage of ammonia converted to nitric acid in each of 50 repetitions of an experiment.
   (e) The number of hours of operation in a given month for 40 nominally identical computers.
   (f) The number of right turns made by 100 rats, each rat having ten encounters with T-junctions in a maze.
   (g) The systolic blood pressure of 80 expectant mothers.
   (h) The number of errors (assume a maximum of 5) found by a firm of auditors in 100 balance sheets.
   (i) The number of each of six types of room found in a large hotel. The types are: single bedded, double bedded, single and double bedded, each with or without a bath.
   (j) The density of ten blocks of compressed carbon dioxide.
   (k) The number of sheep farms on each type of land. The land types are: flat, hilly and mountainous.
   (l) The fluoride content of the public water supply for 100 cities in the UK.

2. For each of the following data sets, summarize by both a tabular and a graphical method:
   (a) The weight of coffee in a sample of 70 jars marked 200 g was recorded as follows:

200.72	201.35	201.42	202.15	201.57	200.79	201.24	203.16	202.31	204.61
201.92	201.74	200.23	201.99	202.31	203.14	200.64	202.71	202.91	202.76
201.08	200.75	202.76	201.35	200.31	201.62	203.27	200.37	200.62	205.72
202.32	201.27	202.07	200.94	201.46	202.06	200.85	201.82	203.47	203.48
200.47	202.83	201.57	202.25	200.87	201.19	203.72	201.75	201.95	203.97
201.42	202.69	201.64	202.16	201.11	201.69	201.27	203.35	201.72	203.06
203.04	201.57	200.71	201.78	202.91	201.72	202.19	201.32	202.47	202.69

   How would your answer change if you knew that the first 35 jars were filled by one machine and the remaining jars were filled by a second machine?

(b) The number of incoming calls at a telephone exchange in each of 50 minutes were:

1	0	1	1	0	0	2	0	0	1
1	0	1	0	0	1	0	1	0	2
0	0	2	4	0	0	1	2	3	0
0	4	3	1	0	1	1	4	0	2
2	5	4	0	2	0	0	3	2	1

(i) Assume first that the numbers are not in time sequence.

(ii) Then assume that the numbers are in time sequence, reading in rows.

(c) The 300 staff in an office block were asked how far they lived from work. The numbers within the 5-mile intervals: less than 5 miles, between 5 and 10 miles, etc., were 60, 100, 75, 45, 15, 4, 1. The manager claims that half the staff live less than 7 miles from work. Can you be more accurate?

(d) The A-level counts of 42 students in higher education were:

3	16	11	6	6	9	7	4	3	9
8	6	12	6	6	5	6	4	3	5
3	4	2	3	6	4	11	4	3	9
5	5	5	6	8	5	5	4	3	12
7	5								

What graph would you draw if you also knew the IQ of each student?

3. Draw a stem and leaf display for the 70 data values in Question 2(a), first rounding each value to 1 dp. Compare the shape of the display with that of the histogram and comment. Now split each level into two, so that 200· is used for leaves of 0 to 4 and 200∗ is used for leaves 5 to 9, and so on. This is an example of a stretched stem and leaf display.

Summarizing data by numerical measures

Let us condense our calculations into a few general abstracts...

You are probably familiar with the idea of an 'average', and with the introduction of modern scientific calculators you may have heard the term 'standard deviation'. Average and standard deviation are examples of numerical measures which we use to summarize data. There are many other such measures. It is the purpose of this chapter to show how we calculate some of these measures, but it is equally important for you to learn when to use a particular measure in a given situation.

4.1 AVERAGES

In this book the word 'average' will be thought of as a vague word meaning 'a middle value' (a single value which in some way represents all the data). It will only take on a definite meaning if we decide that we are referring to the:

(a) sample (arithmetic) mean, or
(b) sample median, or
(c) sample mode, or
(d) some other rigorously defined 'measure of central tendency'.

Averages will be discussed in Sections 4.2, 4.3, 4.4 and 4.5.

4.2 SAMPLE MEAN ($\bar{x}$)

The sample arithmetic mean or simply the sample mean is defined in words as follows:

$$\text{sample mean} = \frac{\text{total of a number of sample values}}{\text{the number of sample values}}.$$

We will distinguish between cases where the data are ungrouped (because we have relatively few sample values) and data which have been grouped into a group frequency distribution table as discussed in Chapter 3. The symbol we use for the sample mean is $\bar{x}$ (refer to Section 2.1 if necessary).

For ungrouped data,

$$\bar{x} = \frac{\Sigma x}{n}.$$

Example Ungrouped data

The heights in centimetres of a sample of five people are: 150, 200, 180, 160, 170. The sample mean height is

$$\bar{x} = \frac{\Sigma x}{n} = \frac{150 + 200 + 180 + 160 + 170}{5} = 172 \text{ cm}.$$

(The answer should not be given to more than three significant figures, i.e. one more significant figure than the raw data.)

For grouped data,

$$\bar{x} = \frac{\Sigma fx}{\Sigma f}.$$

In this formula x refers to the mid-point of a group, f refers to frequency and the Σ sign implies summation over all groups.

Example Grouped data

Table 4.1 Sample mean height using data from Table 3.2

Height (cm)	*Group mid-point* x	*Number of students* f	fx
145.0–154.9	150	3	450
155.0–164.9	160	9	1440
165.0–174.9	170	21	3570
175.0–184.9	180	13	2340
185.0–194.9	190	4	760
		$\Sigma f = 50$	$\Sigma fx = 8560$

The sample mean height is

$$\bar{x} = \frac{\Sigma fx}{\Sigma f} = \frac{8560}{50} = 171.2 \text{ cm.}$$

Notes

(a) Strictly speaking the group mid-points are:

$$\tfrac{1}{2}(145.0 + 154.9) = 149.95, \text{ etc.}$$

but we will not worry about such small discrepancies here.

(b) The fx column is formed by multiplying values in the f column by corresponding values in the x column. The fx values are the total heights of students in each group assuming each has the mid-point height of the group. Σfx means 'sum the fx values'. Thus Σfx is the total height of all 50 students. Since Σf is the total number of students, we see that the formula for $\bar{x}$ is essentially the same as for ungrouped data.

(c) The formula for $\bar{x}$ can be used even if the groups are of unequal width.

(d) If the data are discrete and groups consist of single values of the variable (as in Table 3.4) then these values are also the group mid-points.

(e) If the data are categorical (as in Table 3.5) the sample mean has no meaning and hence is not defined for such data.

4.3 SAMPLE MEDIAN

The sample median is defined as the middle value when the data values are ranked in increasing (or decreasing) order of magnitude.

Example Ungrouped data

The heights in centimetres of a sample of five people are: 150, 200, 180, 160, 170, or, in rank order: 150, 160, 170, 180, 200.

The sample median height = 170 cm.

In general, for ungrouped data consisting of n values arranged in order of magnitude, the median is the $\left(\frac{n+1}{2}\right)$th value.

Example Grouped data

For the sample median of continuous[†] grouped data we will use a graphical method, namely the cumulative frequency polygon (refer to Section 3.6 if

[†] For discrete grouped data, the median can be obtained directly from the grouped frequency distribution table. So for the example of Table 3.4, there are 15 values of 0, 20 values of 1, etc. Since $\left(\frac{n+1}{2}\right) = 25\tfrac{1}{2}$, the median is halfway between the 25th and 26th value, and so must equal 1. Sample median number of siblings = 1. For categorical grouped data the sample median has no meaning and is not defined for such data.

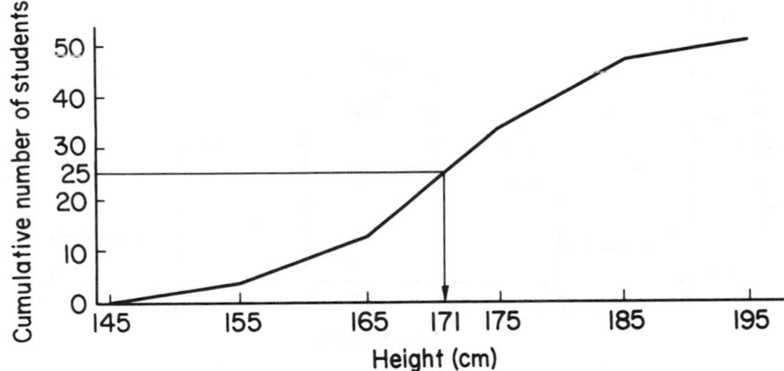

Fig. 4.1 Cumulative frequency polygon for the data in Table 3.3.

necessary). For the data from Table 3.3, we drew Fig. 3.2, reproduced here as Fig. 4.1.

Since the sample median is defined as the middle value when the data values are ranked in increasing order of magnitude, this implies that half the values are less than the sample median. To find the sample median we can use the cumulative frequency polygon by finding the height corresponding to half the total frequency. For the example we read the graph at a cumulative frequency of $1/2 \times 50 = 25$, hence:

$$\text{sample median height} = 171 \text{ cm}.$$

4.4 SAMPLE MODE

The sample mode is defined as the value which occurs with the highest frequency.

Example Ungrouped data

The heights in centimetres of a sample of five people are: 150, 200, 180, 160, 170. All five values occur with a frequency of one, so all have a claim to be the sample mode. The fact that the sample mode may not be a unique value is one of the mode's disadvantages – there are others.

Example Grouped data

For the sample mode of continuous† grouped data we will use a graphical method, namely the histogram (refer to Section 3.6 if necessary). For the data from Table 3.2 we drew Fig. 3.1, reproduced here as Fig. 4.2.

† For both discrete and categorical data the sample mode can be read directly from the grouped frequency distribution table. So for the example of Table 3.4 the modal number of siblings is 1, since this value occurs with the highest frequency. And for the example of Table 3.5, the modal hair colour is black.

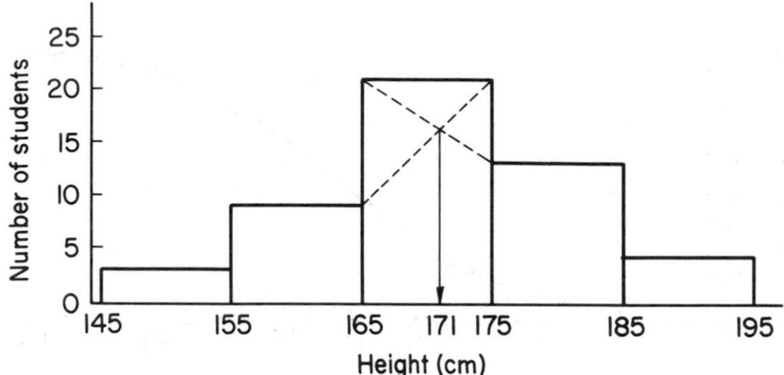

Fig. 4.2 Histogram for the data in Table 3.2.

The modal group is 165–175 since this group has the highest frequency. To find the modal value a cross is drawn as shown in Fig. 4.2. A vertical line is drawn through the point where the arms of the cross meet. The sample mode is the value indicated by where the vertical line meets the horizontal axis.

For this example, sample modal height = 171 cm.

4.5 WHEN TO USE THE MEAN, MEDIAN AND MODE

In order to decide which of the three 'averages' to use in a particular case we need to consider the shape of the distribution as indicated by the histogram (for continuous data) or the line chart (for discrete data). For categorical data the mode is the only average which is defined.

If the shape is roughly symmetrical about a vertical centre line, and is unimodal (single peak), then the sample mean is the preferred average. Such is the case in Fig. 4.2. You may have noticed, however, that the values for the mean, median and mode were almost identical for these data:

sample mean = 171.2, sample median = 171, sample mode = 171.

So why should any one measure be preferred in this case? The answer is a theoretical one, which you are asked to take on trust, that the sample mean is a more precise measurement for such distributions. (Precise here means that, if lots of random samples, each of 50 students, were drawn from the same population of heights, there would be less variation in the sample means than in the sample medians or sample modes of these samples. The terms 'random sample' and 'population' will be discussed in Chapter 8.)

If the shape of the distribution is not symmetrical it is described as

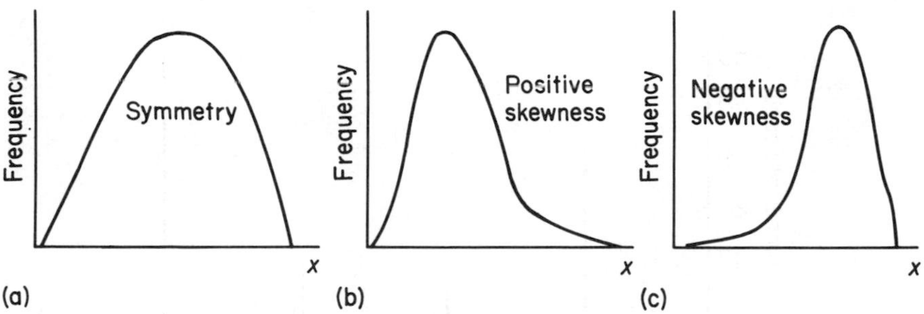

Fig. 4.3 (a) Mean = Median = Mode; (b) Mean > Median > Mode; (c) Mean < Median < Mode.

'skew'. Figs. 4.3(a), (b) and (c) are sketches of unimodal (single peak) distributions which exhibit symmetry, positive skewness and negative skewness, respectively.

Figure 4.3 also indicates the rankings of the sample mean, sample median and sample mode in each of the three cases. For markedly skew data there will be a small number of extremely high values (Fig. 4.3(b)) or low values (Fig. 4.3(c)) which are not balanced by values on the other side of the distribution. The sample mean is more influenced by these few extreme values than the sample median. So the sample median is preferred for data showing marked skewness. By marked skewness we mean that the measure of skewness (see Section 4.12) is greater than 1, or less than −1, as a rough guide. If in doubt both the sample mean and sample median should be quoted.

The mode is not much use for either continuous or discrete data, since it may not be unique (as we have seen in Section 4.4), it may not exist at all, and for other theoretical reasons. The mode is useful only for categorical data.

Occasionally, distributions arise for which none of the three 'averages' is informative. For example, suppose we carry out a survey of the readership of a monthly magazine, noting the number of magazines read by each of a random sample of 100 people over the last three months. Probably most people will have read the magazine every month or not at all as indicated in Fig. 4.4. The sample mean number of magazines per person is

$$\frac{35 \times 0 + 5 \times 1 + 10 \times 2 + 50 \times 3}{100} = 1.75.$$

The sample median is any number between 2 and 3. The sample mode is 3.

The 'U-shaped' graph provides a more meaningful summary than any of the three measures. For example, it indicates that half the sample are regular readers.

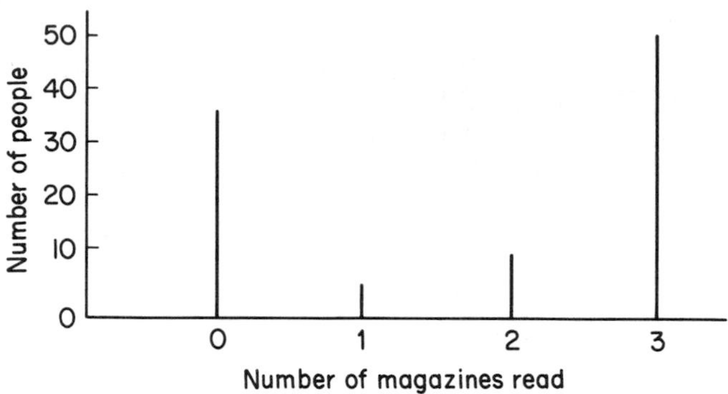

Fig. 4.4 Readership of a monthly magazine over three months.

4.6 MEASURES OF VARIATION

Averages are not the whole story. They do not give a complete description of a set of data and can, on their own, be misleading. So the definition of a statistician as one who, on plunging one foot into a bucket of boiling water and the other into a bucket of melting ice, declares 'On average I feel just right!' completely misses the purpose of statistics which is to collect and analyse data which vary. It is not the aim of this book to lament the misconceptions some people have about statistics, but hopefully to inform and educate. So it would be more reasonable for the caricatured statistician to feel unhappy because the temperature of his feet vary so greatly about a comfortable average. Similarly an employee feels unhappy when told, 'wages have risen by 10% in the past year', if his wage has risen by only 3%, while the cost of living has risen by 8% (both the 10% and the 8% are averages, by the way).

Two measures of variation will be discussed in Sections 4.7, 4.8 and 4.9 in some detail, and three other measures of variation will be mentioned briefly in Section 4.10.

4.7 SAMPLE STANDARD DEVIATION(s)

One way of measuring variation in sample data is to sum the differences between each value and the sample mean, $\bar{x}$, to give:

$$\Sigma(x - \bar{x}).$$

However, this always gives zero, as we saw in Chapter 2.

A more useful measure of variation, called the sample standard deviation, s, is obtained by summing the squares of the differences $(x - \bar{x})$, dividing by $n - 1$, and then taking the square root. This gives a kind of 'root mean square deviation' (see the formula for s below).

The reason for squaring the differences is that this makes them all positive or zero. The reason for dividing by $n - 1$ rather than n is discussed later in this section. The reason for taking the square root is to make the measure have the same units as the variable x. There are more theoretical reasons than these for using standard deviation as a measure of variation, but I hope the above will give you an intuitive feel for the formulae which are now introduced.

For ungrouped data, the sample standard deviation, s, may be defined by the formula:

$$s = \sqrt{\frac{\Sigma(x - \bar{x})^2}{n - 1}},$$

but an alternative form of this formula which is easier to use for calculation purposes is

$$s = \sqrt{\frac{\Sigma x^2 - \frac{(\Sigma x)^2}{n}}{n - 1}}.$$

Example Ungrouped data

The heights in centimetres of a sample of five people are: 150, 200, 180, 160, 170.

$$\Sigma x = 150 + 200 + 180 + 160 + 170 = 860$$
$$\Sigma x^2 = 150^2 + 200^2 + 180^2 + 160^2 + 170^2 = 149\,400.$$

The sample standard deviation is

$$s = \sqrt{\frac{149\,400 - \frac{860^2}{5}}{4}} = 19.2 \text{ cm}.$$

Notes

(a) The answer should not be given to more than three significant figures, i.e. one more than the raw data.

(b) The units of standard deviation are the same as the units of the variable height, i.e. centimetres.

The formula for the standard deviation for grouped data is

$$s = \sqrt{\frac{\Sigma fx^2 - \frac{(\Sigma fx)^2}{\Sigma f}}{\Sigma f - 1}}.$$

Example Grouped data

Table 4.2 Sample standard deviation of data from Table 3.2

Height (cm)	*Group mid-point* x	*Number of students* f	fx	fx^2
145.0–154.9	150	3	450	67 500
155.0–164.9	160	9	1440	230 400
165.0–174.9	170	21	3570	606 900
175.0–184.9	180	13	2340	421 200
185.0–194.9	190	4	760	144 400
		$\Sigma f = 50$	$\Sigma fx = 8560$	$\Sigma fx^2 = 1\,470\,400$

The sample standard deviation is

$$s = \sqrt{\frac{1\,470\,400 - \dfrac{8560^2}{50}}{49}} = 10.03 \text{ cm.}$$

Notes

(a) This table is the same as Table 4.1 except for the extra fx^2 column, which may be formed by multiplying the values in the x column by corresponding values in the fx column. Usually we will calculate both the sample mean and sample standard deviation from a table such as Table 4.2.

(b) The formula for s can be used even if the groups are of unequal width.

A common question which is often asked is, 'Why use $(n - 1)$ and similarly $\Sigma f - 1$ in the formulae for s?' The answer is that the values we obtain give better estimates of the standard deviation of the population that would be obtained if we used n (and similarly Σf) instead. In what is called statistical inference (Chapter 8 onwards) we are interested, not so much in sample data, but in what conclusions, based on sample data, can be drawn about a population.

Another natural question at this stage is, 'Now we have calculated the sample standard deviation, what does it tell us?' The answer is 'be patient!'. When we have discussed the normal distribution in Chapter 7, standard deviation will become more meaningful. For the moment please accept the basic idea that standard deviation is a measure of variation about the mean. The more variation in the data, the higher will be the standard deviation. If there is no variation at all, the standard deviation will be zero. It can never be negative.

4.8 SAMPLE INTER-QUARTILE RANGE

Just as the sample median is such that half the sample values are less than it, the lower quartile and the upper quartile are defined as follows:

The *lower quartile* is the value such that one-quarter of the sample values are less than it.
The *upper quartile* is the value such that three-quarters of the sample values are less than it.

The *sample inter-quartile range* is then defined by the formula

$$\text{sample inter-quartile range} = \text{upper quartile} - \text{lower quartile}$$

Example Ungrouped data

The heights in centimetres of a sample of five people are: 150, 200, 180, 160, 170, or, in rank order: 150, 160, 170, 180, 200.

For ungrouped data consisting of n values arranged in order of magnitude, the lower and upper quartiles are the $\frac{1}{4}(n + 1)$th and $\frac{3}{4}(n + 1)$th values respectively. Here $n = 5$, the lower quartile is $\frac{1}{2}(150 + 160) = 155$, the upper quartile is $\frac{1}{2}(180 + 200)$, giving an inter-quartile range of $190 - 155 = 35$.

Example Grouped data

As for the median, we use the cumulative frequency polygon to obtain the sample inter-quartile range of continuous† grouped data.

For the data from Table 3.3 we drew Fig. 3.2, reproduced here as Fig. 4.5.

The lower quartile is the height corresponding to one-quarter of the total frequency, the upper quartile is the height corresponding to three-quarters of the total frequency. For the example read the graph at cumulative frequencies of $(\frac{1}{4}) \times 50 = 12.5$ and $(\frac{3}{4}) \times 50 = 37.5$, hence lower quartile $= 165$ cm, upper quartile $= 178$ cm, and sample inter-quartile range $= 178 - 165 = 13$ cm. Notice that the 'middle half' of the sample values lie between the lower and upper quartiles.

4.9 WHEN TO USE STANDARD DEVIATION AND INTER-QUARTILE RANGE

In order to decide which of these measures of variation to use in a particular case the same considerations apply as for averages (refer to Section 4.5 if

† For discrete data, the sample inter-quartile range can be obtained directly from the grouped frequency distribution (in a similar way to the median). You should check for yourself that for the example in Table 3.4, the lower quartile is 0, the upper quartile is 2, so the sample inter-quartile range = 2. For categorical data, the sample inter-quartile range has no meaning and is not defined for such data.

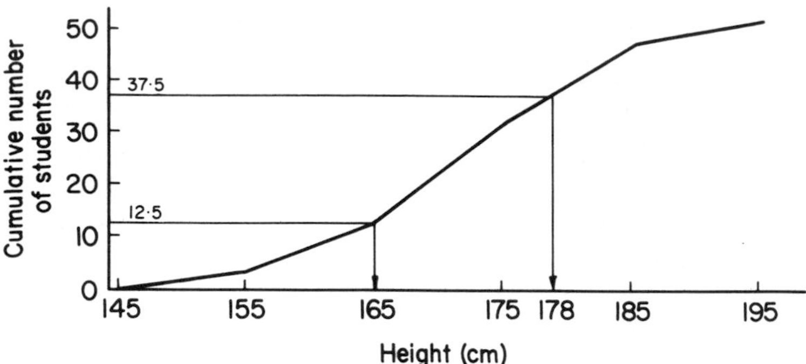

Fig. 4.5 Cumulative frequency polygon for the data in Table 3.3.

necessary). For roughly symmetrical data, use standard deviation. For markedly skew data, use sample inter-quartile range.

4.10 OTHER MEASURES OF VARIATION

We will consider three other measures of variation briefly:

(a) *Variance* is simply the square of the standard deviation, so we can use the symbol s^2. Variance is a common term in statistical methods beyond the scope of this book, for example in 'analysis of variance'.
(b) *Coefficient of variation* is defined as $100s/\bar{x}$ and is expressed as a percentage. This is used to compare the variability in two sets of data when there is an obvious difference in magnitude in both the means and standard deviations. For example to compare the variation in heights of boys aged 5 and 15, suppose $\bar{x}_5 = 100$, $s_5 = 6$, $\bar{x}_{15} = 150$, $s_{15} = 9$, then both sets have a coefficient of variation of 6%.
(c) *Range* is defined as (largest value − smallest value). It is commonly used because it is simple to calculate, but it is unreliable, except in special circumstances, because only two of the sample values are used to calculate it. Also the more sample values we take, the larger the range is likely to be.

4.11 BOX AND WHISKER PLOTS

These plots come under the heading of exploratory data analysis (see Section 3.10), and are a graphical method of representing five statistical summaries of a set of univariate data, namely:

the smallest value, the lower quartile, the median, the upper quartile and the largest value.

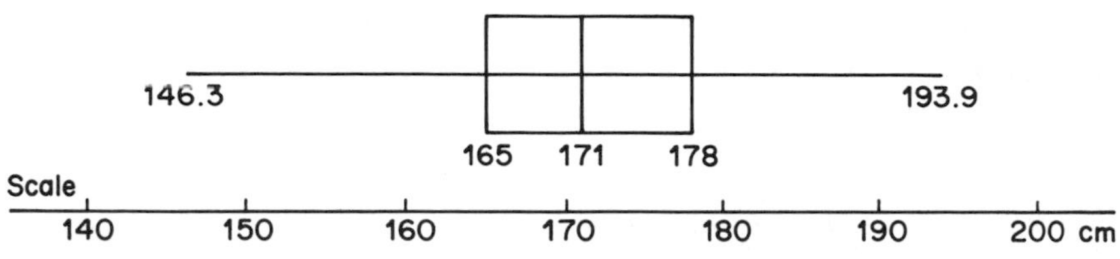

Fig. 4.6 A box and whisker plot for the data in Table 3.2

Example

Draw a box and whisker plot for the data in Table 3.1. For these data, we have already found the following:

smallest value	= 146.3 cm	(Section 3.1)
lower quartile	= 165 cm	(Section 4.8)
median	= 171 cm	(Section 4.3)
upper quartile	= 178 cm	(Section 4.8)
largest value	= 193.9 cm	(Section 3.1)

It is clear from Fig. 4.6 that the box ends represent the quartiles, that the box is divided by the median into two (usually unequal) parts, and that the whiskers are the lines joining the smallest and largest values to the lower and upper quartiles respectively.

4.12 MEASURES OF SHAPE

As well as averages and measures of variation we will consider two measures of skewness (referred to already in Section 4.5), namely

$$\frac{\text{sample mean} - \text{sample mode}}{\text{sample standard deviation}} \quad \text{and} \quad \frac{3(\text{sample mean} - \text{sample median})}{\text{sample standard deviation}}.$$

These measures give approximately the same value for unimodal data (data for which the histogram has one peak). For perfectly symmetrical data the measures of skewness will be zero. They will be positive for positively skew data and negative for negatively skew data.

For the second of the measures, the possible range of values is -3 to $+3$.

4.13 SUMMARY

When a variable is measured for a number of individuals the resulting data may be summarized by calculating averages and measures of variation. In addition a measure of shape is sometimes useful.

The particular type of average and measure of variation depends on the type of variable and the shape of the distribution. The formula or method of calculating the measure also depends on whether the data are grouped or ungrouped.

Some examples are:

Type of variable	*Shape of distribution*	*Average*	*Measure of variation*
Continuous or discrete	Roughly symmetrical, unimodal	Sample mean ($\bar{x}$)	Sample standard deviation(s)
Continuous or discrete	Markedly skew,	Sample median	Sample inter-quartile range
Categorical	Unimodal	Sample mode	

Three other measures of variation are variance, coefficient of variation and range.

WORKSHEET 4: SUMMARIZING DATA BY NUMERICAL MEASURES

1. (a) Why do we need averages?
   (b) Which average can have more than one value?
   (c) Which average represents the value when the total of all the sample values is shared out equally?
   (d) Which average has the same number of values above it as below it?
   (e) When is the sample median preferred to the sample mean?
   (f) When is the sample mode preferred to the sample mean?
   (g) When is the sample mean preferred to both the sample median and the sample mode?

2. (a) Why do we need measures of variation?
   (b) What measure of variation is most useful in the case of: (i) a symmetrical distribution, (ii) a skew distribution?
   (c) Think of an example of sample data where the range would be a misleading measure of variation.
   (d) Name the measure of variation associated with the: (i) sample mean ($\bar{x}$), (ii) sample median, (iii) sample mode.
   (e) Name the average associated with the: (i) sample standard deviation, (ii) sample inter-quartile range, (iii) range.

3. The weekly incomes (£) of a random sample of self-employed window cleaners are: 75, 67, 60, 62, 65, 67, 62, 68, 82, 67, 62, 200. Do not group these data.
   (a) Find the sample mean, sample median and sample mode of weekly income. Why are your answers different?

(b) Find the sample standard deviation and sample inter-quartile range of weekly income. Why are your answers different?
(c) Which of the measures you have obtained are the most useful in summarizing the data?

4. Eleven cartons of sugar, each nominally containing 1 kg, yielded the following weights of sugar:

   1.02, 1.05, 1.08, 1.03, 1.00, 1.06, 1.08, 1.01, 1.04, 1.07, 1.00.

   Calculate the sample mean and sample standard deviation of the weight of sugar.
   Note If you do not have a facility for standard deviation on your calculator you may be tempted to round the value of Σx^2 and $(\Sigma x)^2$ to fewer than 4 dps before finally calculating s. This could lead to grossly inaccurate answers. Be warned!

5. Using the data from Worksheet 3, Question 2(a):
   (a) Find the sample mean and sample standard deviation of the variable.
   (b) Find the sample median and sample inter-quartile range of the variable.
   (c) Which of the measures you have obtained are the most useful in summarizing the data?

6. Repeat Question 5 for the following data which are the 'distances travelled' in thousands of kilometres by 60 tyres which were used in simulated wear tests before they reached the critical minimum tread.

Distance (km × 1000)	*Number of tyres*
From 16 up to but not including 24	5
From 24 up to but not including 32	10
From 32 up to but not including 40	20
From 40 up to but not including 48	16
From 48 up to but not including 56	9

7. Repeat Question 5 for the data in Worksheet 3, Question 2(b).

8. Repeat Question 5 for the data in Worksheet 3, Question 2(c).

9. Repeat Question 5 for the data in Worksheet 3, Question 2(d).

10. Draw a box and whisker plot for the data in Worksheet 3, Question 2(a), using the results of Question 5(b) above.

Note

In this chapter we have seen a tabular method of obtaining $\bar{x}$ and s for grouped data. With some scientific calculators it is possible to obtain $\bar{x}$ and s by typing in only the group mid-points and the corresponding frequencies. This may be performed in SD mode by entering, for each group in turn, mid-point × frequency M+, and then pressing the mean and standard deviation buttons.

Probability

> Dr Price estimates the chance in favour of the wife being the survivor in marriage as 3 to 2.

5.1 INTRODUCTION

The opening chapters of this book have been concerned with statistical data and methods of summarizing such data. We can think of such sample data as having been drawn from a larger 'parent' population. Conclusions from sample data about the populations (which is a branch of statistics called 'statistical inference'; see Chapter 8 onwards) must necessarily be made with some uncertainty since the sample cannot contain all the information in the population.

This is one of the main reasons why probability, which is a measure of uncertainty, is now discussed.

Probability is a topic which may worry you, either if you have never studied it before, or if you have studied it before but you did not fully get to grips with it. It is true that the study of probability requires a clear head, a logical approach and the ability to list all the outcomes of simple experiments often with the aid of diagrams. After some experience and some possibly painful mistakes, which are all part of the learning process, the penny usually begins to drop.

Think about the following question which will give you some feedback on your present attitude towards probability (try not to read the discussion until you have thought of an answer).

Probability Example 5.1

A person tosses a coin five times. Each time it comes down heads. What is the probability that it will come down heads on the sixth toss?

Discussion

If your answer is '1/2' you are assuming that the coin is 'fair', meaning that it is equally likely to come down heads or tails. You have ignored the data that all five tosses resulted in heads.

If your answer is 'less than 1/2' you may be quoting 'the law of averages' which presumably implies that, in the long run, half the tosses will result in heads and half in tails. This again assumes that the coin is fair. Also, do six tosses constitute a long run of tosses, and does the 'law of averages' apply to each individual toss?

If your answer is 'greater than 1/2', perhaps you suspect that the coin has two heads, in which case the probability of heads would be 1, or that the coin has a bias in favour of heads. Think about this teasing question again when you have read this chapter.

5.2 BASIC IDEAS OF PROBABILITY

One dictionary definition of probability is 'the extent to which an event is likely to occur, measured by the ratio of the favourable cases to the whole number of cases possible'. Consider the following example.

Probability Example 5.2

'From a bag containing three red balls and seven white balls, the probability of a red ball being drawn first is 3/10.'

The main idea to grasp at this stage is that probability is a measure. However, in order to gain a better understanding of probability, we need to consider three terms which are in everyday use, but which have specific meanings when we discuss probability, namely trial, experiment and event. These terms are defined as follows:

A *trial* is an action which results in one of several possible outcomes.
An *experiment* consists of a number of trials.
An *event* is a set of outcomes which have something in common.

For Example 5.2 above, the experiment consists of only one trial, namely drawing a ball from the bag. The event, 'red ball', corresponds to the set of three outcomes, s[illegible] there are three red balls in the bag.

5.3 THE *A PRIORI* DEFINITION OF PROBABILITY FOR EQUALLY LIKELY OUTCOMES

Suppose each trial in an experiment can result in one of n 'equally likely' outcomes, r of which correspond to an event, E, then the probability of event E is r/n, which we write

$$P(E) = \frac{r}{n}.$$

This *a priori* definition has been used for Example 5.2; event E is 'red ball', $n = 10$ since it is assumed that each of the 10 balls is equally likely to be drawn from the bag, and $r = 3$ since 3 of the 10 balls in the bag are red and therefore correspond to the event E. So we write

$$P(\text{red ball}) = \frac{3}{10}.$$

Note the following points:

(a) We only have to think about the possible outcomes, we do not have to actually carry out an experiment of removing balls from the bag. The Latin phrase *a priori* means 'without investigation or sensory experience'.
(b) It is necessary to know that the possible outcomes are equally likely to occur. This is why this definition of probability is called a 'circular' definition, since equally likely and equally probable have the same meaning. More importantly we should not use the *a priori* definition if we do not know that the possible outcomes are equally likely. (Example: 'Either I will become Prime Minister or I will not, so the probability that I will is 1/2, and the same probability applies to everyone'. This is clearly an absurd argument.)

The *a priori* definition is most useful in games of chance.

5.4 THE RELATIVE FREQUENCY DEFINITION OF PROBABILITY, BASED ON EXPERIMENTAL DATA

Probability Example 5.3

Suppose we consider all the couples who married in 1970 in the UK. What is the probability that, if we select one such couple at random, they will still be married to each other today? We have no reason to assume that the outcomes 'still married' and 'not still married' are equally likely. We need to look at the experimental data, that is to carry out a survey to find out what state those marriages are in today, and the required probability would be the ratio, i.e. the relative frequency, given by:

$$\frac{\text{number of couples still married today who married in 1970 in the UK}}{\text{number of couples who married in 1970 in the UK}}.$$

We will not consider the practical problems of gathering such data!

Formally, the relative frequency definition of probability is as follows: If in a large number of trials, n, r of these trials result in event E, the probability of event E is r/n. So we write

$$P(E) = \frac{r}{n}.$$

Notes

(a) n must be large. The larger the value of n, the better is the estimate of probability.
(b) The phrase 'at random' above means that each couple has the same chance of being selected from the particular population.
(c) One theoretical problem with this definition of probability is that apparently there is no guarantee that the value of r/n will settle down to a constant value as the number of trials gets larger and larger. Luckily this is not a practical problem, so we shall not pursue it.

5.5 THE RANGE OF POSSIBLE VALUES FOR A PROBABILITY VALUE

Using either of the two definitions of probability, we can easily show that probability values can only be between 0 and 1. The value of r must take one of the integer values between 0 and n, so r/n can take values between $0/n$ and n/n, that is 0 and 1.

If $r = 0$ we are thinking of an event which cannot occur (*a priori* definition) or an event which has not occurred in a large number of trials (relative frequency definition). For example, the probability that we will throw a 7 with an ordinary die is 0. If $r = n$ we are thinking of an event which must always occur (*a priori* definition) or an event which has occurred in each of a large number of trials (relative frequency definition). For example, the probability that the sun will rise tomorrow can be assumed to be 1, unless you are a pessimist (see Section 5.7).

5.6 PROBABILITY, PERCENTAGE, PROPORTION AND ODDS

We can convert a probability to a percentage by multiplying it by 100. So a probability of 3/4 implies a percentage of 75%.

We can also think of probability as meaning the same thing as proportion. So a probability of 3/4 implies that the proportion of times the event will occur is also 3/4.

A probability of 3/4 is equivalent to odds of 3/4 to 1/4, which is usually expressed as 3 to 1.

Note
You are advised to try Worksheet 5, Questions 1–7 before proceeding with this chapter.

5.7 SUBJECTIVE PROBABILITY

There are other definitions of probability apart from the two discussed earlier in this chapter. We all use 'subjective probability' in forecasting future events, for example when we try to decide whether it will rain tomorrow, and when we try to assess the reaction of others to our opinions and actions. We may not be quite so calculating as to estimate a probability value, but we may regard future events as being probable, rather than just possible. In subjective assessments of probability we may take into account experimental data from past events, but we are likely to add a dose of subjectivity depending on our personality, our mood and other factors.

5.8 PROBABILITIES INVOLVING MORE THAN ONE EVENT

Suppose we are interested in the probabilities of two possible events, E_1 and E_2. For example, we may wish to know the probability that both events will occur, or the probability that either or both events will occur. We will refer to these as, respectively:

$$P(E_1 \text{ and } E_2) \qquad \text{and} \qquad P(E_1 \text{ or } E_2 \text{ or both}).$$

In set theory notation these compound events are called the intersection and union of events E_1 and E_2 respectively, and their probabilities are written:

$$P(E_1 \cap E_2) \qquad \text{and} \qquad P(E_1 \cup E_2).$$

There are two probability laws which can be used to estimate such probabilities, and these are discussed in Sections 5.9 and 5.10.

5.9 MULTIPLICATION LAW (THE 'AND' LAW)

The general case of the multiplication law is:

$$P(E_1 \text{ and } E_2) = P(E_1)P(E_2|E_1)$$

where $P(E_2|E_1)$ means the probability that event E_2 will occur, given that event E_1 has already occurred. The vertical line between E_2 and E_1 should be read as 'given that' or 'on the condition that'.

$P(E_2|E_1)$ is an example of what is called a *conditional* probability.

Probability Example 5.4

If two cards are selected at random, one at a time without replacement from a pack of 52 playing cards, what is the probability that both cards will be aces?

$$\text{P(two aces)} = \text{P(first card is an ace and second card is an ace)},$$

the two phrases in brackets being alternative ways of describing the same result. Using the multiplication law, the right-hand side can now be written as:

$$\text{P(first card is an ace)} \times \text{P(second card is an ace}|\text{first card is an ace)}$$

$$= \frac{4}{52} \times \frac{3}{51},$$

using the *a priori* definition, and taking into account the phrase 'without replacement' above

$$= 0.0045.$$

(4 dps are usually more than sufficient for a probability value.)

In many practical examples the probability of event E_2 does not depend on whether E_1 has occurred. In this case we say that events E_1 and E_2 are *statistically independent*, giving rise to the special case of the multiplication law:

$$\text{P}(E_1 \text{ and } E_2) = \text{P}(E_1)\text{P}(E_2).$$

Probability Example 5.5

If two cards are selected at random, one at a time with replacement (so the first card is replaced in the pack before the second is selected) from a pack of 52 playing cards, what is the probability that both cards will be aces?

$$\begin{aligned}\text{P(two aces)} &= \text{P(first card is an ace and second card is a ace)} \\ &= \text{P(first card is an ace)} \times \text{P(second card is an ace)} \\ &= \frac{4}{52} \times \frac{4}{52} \\ &= 0.0059.\end{aligned}$$

Because the first card is replaced, the probability that the second card is an ace will be 4/52 whatever the first card is. We have therefore used the special case of the multiplication law because the events 'first card is an ace' and 'second card is an ace' are statistically independent.

Notes

(a) Comparing the two cases of the multiplication law we can state that, if two events E_1 and E_2 are statistically independent,

$$P(E_2|E_1) = P(E_2).$$

(b) Clearly we could write the general law of multiplication alternatively as $P(E_1 \text{ and } E_2) = P(E_2)P(E_1|E_2)$.

5.10 ADDITION LAW (THE 'OR' LAW)

The general case of the addition law is:

$$P(E_1 \text{ or } E_2 \text{ or both}) = P(E_1) + P(E_2) - P(E_1 \text{ and } E_2).$$

Probability Example 5.6

If a die is tossed twice, what is the probability of getting at least one 6? This is the same as asking for

$$\begin{aligned} &P(6 \text{ on first toss or } 6 \text{ on second toss or } 6 \text{ on both tosses}), \\ &= P(6 \text{ on first toss}) + P(6 \text{ on second toss}) - P(6 \text{ on both tosses}), \\ &= \frac{1}{6} + \frac{1}{6} - \frac{1}{6} \times \frac{1}{6}, \\ &= \frac{6}{36} + \frac{6}{36} - \frac{1}{36} \\ &= \frac{11}{36}. \end{aligned}$$

Here E_1 is the event '6 on first toss', E_2 is the event '6 on second toss'. For P(6 on both tosses) we have used the multiplication law for independent events.

In many practical cases the events E_1 and E_2 are such that they cannot both occur. In this case we say that events E_1 and E_2 are *mutually exclusive*, giving rise to the special case of the addition law:

$$P(E_1 \text{ or } E_2) = P(E_1) + P(E_2).$$

Note

Comparing the two versions of the addition law, we see that if E_1 and E_2 are mutually exclusive, $P(E_1 \text{ and } E_2) = 0$.

Probability Example 5.7

If a die is tossed twice, what is the probability that the total score is 11?

This time E_1 is '6 on first toss and 5 on second toss', E_2 is '5 on first toss and 6 on second toss'.

These are the only two ways of obtaining a total score of 11. Since E_1 and E_2 are mutually exclusive (since if E_1 occurs E_2 cannot, and vice versa), we have:

$$\begin{aligned} P(11) &= P(E_1 \text{ or } E_2) \\ &= P(E_1) + P(E_2) \\ &= \frac{1}{6} \times \frac{1}{6} + \frac{1}{6} \times \frac{1}{6} \\ &= \frac{1}{18}. \end{aligned}$$

Here we have used the law of addition for the mutually exclusive events E_1 and E_2, followed by the multiplication law twice for independent events.

5.11 MUTUALLY EXCLUSIVE AND EXHAUSTIVE EVENTS

If all possible outcomes of an experiment are formed into a set of mutually exclusive events, we say they form a *mutually exclusive and exhaustive* set of events which we will call $E_1, E_2, \ldots, E_n$, if there are n events. Applying the special case of the law of addition, $P(E_1 \text{ or } E_2 \text{ or} \ldots \text{or } E_n) = P(E_1) + P(E_2) + \cdots + P(E_n)$. But since the events are exhaustive, one of them must occur, and so the left-hand side is 1. In words, then, this result is:

> The sum of the probabilities of a set of mutually exclusive and exhaustive events is 1.

This result is useful in checking whether we have correctly calculated the separate probabilities of the various mutually exclusive events of an experiment. It is also a result for which we will find an application in Chapter 6.

Probability Example 5.8

For families with two children, what are the probabilities of the various possibilities, assuming boys and girls are equally likely at each birth?

Four mutually exclusive and exhaustive events are BB, BG, GB and GG where, for example, BG means a boy followed by a girl.

$$P(BG) = \frac{1}{2} \times \frac{1}{2} = \frac{1}{4},$$

using the special law of multiplication. Similarly P(BB) = P(GB) = P(GG) = 1/4, and the total probability is 1.

5.12 COMPLEMENTARY EVENTS AND THE CALCULATION OF P(AT LEAST 1...)

For any event E, there is a *complementary* event E', which we call 'not E'. Since either E or E' must occur, and they cannot both occur,

$$P(E) + P(E') = 1,$$

a special case of the result of the previous section. So $P(E) = 1 - P(E')$.

This is useful in some calculations where we will find that it is easier to calculate the probability of the complement to some event and subtract the answer from 1, than it is to calculate directly the probability of the event. This is especially true when we wish to calculate the probability that 'at least 1 of something will occur in a number of trials', since

$$P(\text{at least } 1\ldots) = 1 - P(\text{none}\ldots).$$

Probability Example 5.9

For families with four children, what is the probability that there will be at least one boy, assuming boys and girls are equally likely?

Instead of listing the 16 outcomes BBBB, BBBG, etc., we simply use:

$$\begin{aligned} P(\text{at least 1 boy}) &= 1 - P(\text{no boys}) \\ &= 1 - P(\text{GGGG}) \\ &= 1 - \frac{1}{2} \times \frac{1}{2} \times \frac{1}{2} \times \frac{1}{2} \\ &= \frac{15}{16}. \end{aligned}$$

5.13 PROBABILITY TREES

The probability tree (sometimes called a tree diagram) can be used instead of the laws of probability when we are considering the outcomes of a sequence of (a few) trials.

Probability Example 5.10

If two cards are selected at random, one at a time without replacement, from a pack of 52 playing cards, investigate the events ace (A) and not ace (A'). See Fig. 5.1.

In general, the events resulting from the first trial are represented by two or more branches from a starting point. More branches are added to represent the events resulting from the second and subsequent trials. The branches are labelled with the names of the events and their probabilities,

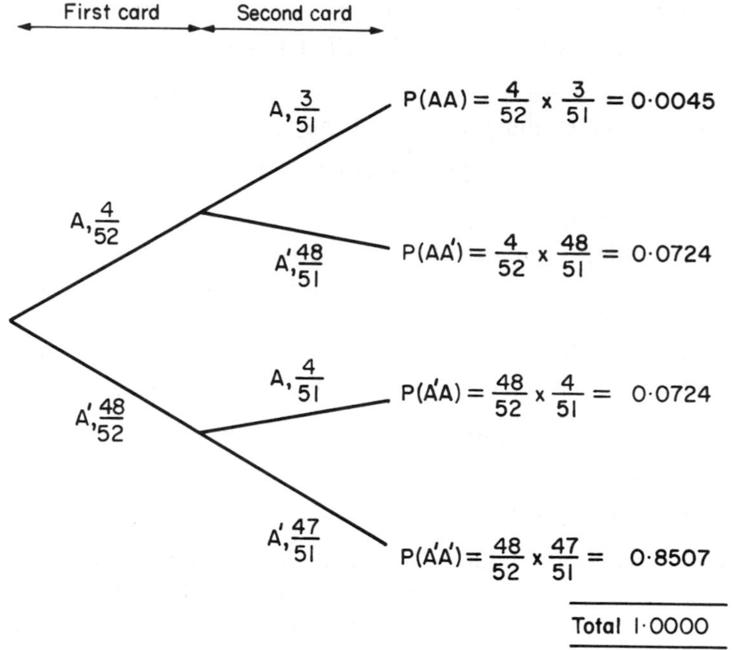

Fig. 5.1 Probability tree for two cards without replacement.

taking into account previous branches (so we might be dealing with conditional probabilities as in Example 5.4). To calculate the probabilities of the set of mutually exclusive and exhaustive events, corresponding to all the ways of getting from the starting point to the end of a branch, probabilities are multiplied (so we are really using the law of multiplication). The total probability should equal 1.

For the example above:

$$\text{P(two aces)} = \text{P(AA)} = 0.0045$$

$$\text{P(one ace)} = \text{P(AA}' \text{ or A}'\text{A)} = \text{P(AA}') + \text{P(A}'\text{A)} = 0.1448$$

$$\text{P(no aces)} = \text{P(A}'\text{A}') = 0.8507.$$

In general, the total number of branches needed will depend on both the number of possible events resulting from each trial and the number of trials. For example, if each of four trials can result in one of two possible events there will be $2 + 4 + 8 + 16 = 30$ branches, or if each of three trials can result in one of three possible events there will be $3 + 9 + 27 = 39$ branches.

Clearly it becomes impractical to draw the probability tree if the number of branches is large. In the next chapter we will discuss more powerful ways of dealing with any number of independent trials each with two possible outcomes.

5.14 SUMMARY

Probability as a measure of uncertainty is defined using the *a priori* and 'relative frequency' definitions. The first is useful in games of chance, the second when we have sufficient experimental data.

In calculating probabilities involving more than one event, two laws of probability are useful:

1. The multiplication law $P(E_1 \text{ and } E_2) = P(E_1)P(E_2|E_1)$, which reduces to $P(E_1 \text{ and } E_2) = P(E_1)P(E_2)$ for statistically independent events.
2. The addition law $P(E_1 \text{ or } E_2 \text{ or both}) = P(E_1) + P(E_2) - P(E_1 \text{ and } E_2)$, which reduces to $P(E_1 \text{ or } E_2) = P(E_1) + P(E_2)$ for mutually exclusive events.

The probability tree is a useful graphical way of representing the events and probabilities for small experiments.

WORKSHEET 5: PROBABILITY

Questions 1–7 are based on Sections 5.1 to 5.6.

1. Distinguish between the *a priori* and relative frequency definitions of probability.
2. If the probability of a successful outcome of an experiment is 0.2, what is the probability of a failure?
3. When two coins are tossed the result can be two heads, one head and one tail, or two tails, and hence each of these events has a probability of 1/3. What is wrong with this argument? What is the correct argument?
4. A coin is tossed five times. Each time it comes down heads. Hence the probability of heads is 5/5 = 1. Discuss.
5. Three ordinary dice, one yellow, one blue and one green, are placed in a bag. A trial involves selecting one die at random from the bag and rolling it, the colour and score being noted.
   (a) What does 'at random' mean here?
   (b) Write down the set of all possible outcomes (called the 'sample space').
   (c) Are the outcomes equally likely?
   (d) What is the probability of each outcome?
   (e) What are the probabilities of the following events:
      (i) Yellow with any score?

(ii) Yellow with an even score?
(iii) Even score with any colour?
(iv) Yellow 1 or blue 2 or green 3?
(v) Neither even blue nor odd yellow?

6. If one student is selected at random from the 50 students whose height data are represented in the histogram, Fig. 3.1, what is the probability that this height will be between 175 and 185 cm? Also express your answer in terms of the ratio of two areas in Fig. 3.1. What percentage of these students have heights:
(a) Between 175 and 185 cm?
(b) Between 145 and 195 cm?

7. If an ordinary drawing-pin is tossed in the air it can fall in one of two ways: point upwards, which we will call event U, or point downwards, which we will call U′. As it is not possible to obtain an *a priori* estimate of the probability that the drawing pin will fall point upwards, P(U), we can estimate this probability by carrying out an experiment as follows: Toss a drawing-pin 50 times and record the result of each of the 50 trials as U or U′. Use the relative frequency definition to estimate P(U) after 1, 2, 3, 4, 5, 10, 20, 30, 40 and 50 trials. Plot a graph of P(U) against U. It should indicate that the estimated probability fluctuates less as more tosses are performed.

Questions 8–21 are based on Sections 5.8 to 5.14.

8. Write down the following events in symbol form, where A and B are two events: (a) not A, (b) A given B, (c) B given A.

9. What is meant by: (a) P(A|B), (b) P(B|A), (c) P(A′), (d) A and B are statistically independent, (e) A and B are mutually exclusive? For (d) and (e), think of examples.

10. What is the 'and' law of probability, as applied to events A and B? What happens if A and B are statistically independent?

11. What is the 'or' law of probability, as applied to events A and B? What happens if A and B are mutually exclusive?

12. What can be concluded if, (a) P(A|B) = P(A), (b) P(A and B) = 0?

13. What is the probability of a 3 or a 6 with one throw of a die?

14. What is the probability of a red card, a picture card (ace, king, queen or jack), or both, when a card is drawn at random from a pack?

15. A coin is tossed three times. Before each toss a subject guesses the result as 'heads' or 'tails'. If the subject guesses 'tails', what is the probability that he will be correct: (a) three times, (b) twice, (c) once, (d) no times? *Hint*: draw a probability tree.

16. Three marksmen have probabilities 1/2, 1/3 and 1/4 of hitting a target with each shot. If all three marksmen fire simultaneously calculate the probability that at least one will hit the target. (Refer to Probability Example 5.9 if necessary.)

17. 3% of the sparking plugs manufactured by a firm are defective. In a random sample of four plugs, what is the probability that exactly one will be defective?

18. Suppose that of a group of people, 30% own both a house and a car, 40% own a house and 70% own a car. What proportion, (a) own at least a house or a car, (b) of car owners are also house-holders?

19. Of 14 double-bedded rooms in a hotel, 9 have a bathroom. Of 6 single-bedded rooms, 2 have a bathroom.
    (a) What is the probability that, if a room is randomly selected, it will have a bathroom?
    (b) If a room is selected from those with a bathroom, what is the probability that it will be a single room?

20. It is known from past experience of carrying out surveys in a certain area that 25% of the houses will be unoccupied on any given day. In a proposed survey it is planned that houses unoccupied on the first visit will be visited a second time. One member of the survey team calculates that the proportion of houses occupied on either the first or second visit is $1 - (\frac{1}{4})^2 = 15/16$. Another member of the team calculates this proportion as $1 - 2 \times 1/4 = 1/2$.

    Show that both members' arguments are incorrect. Give the correct argument.

21. A two-stage rocket is to be launched on a space mission. The probability that the lift-off will be a failure is 0.1. If the lift-off is successful the probability that the separation of the stages will be a failure is 0.05. If the separation is successful, the probability that the second stage will fail to complete the mission is 0.03.

    What is the probability that the whole mission will: (a) be a success, (b) be a failure?

Discrete probability distributions

6.1 INTRODUCTION

If a discrete variable can take values with associated probabilities it is called a discrete random variable. The values and the probabilities are said to form a *discrete probability distribution*.

For example, the discrete probability distribution for the variable number of heads resulting from tossing a fair coin three times may be represented as in Table 6.1.

Table 6.1 Probability distribution for the number of heads in three tosses of a coin

Number of heads	0	1	2	3
Probability	0.125	0.375	0.375	0.125

The above probabilities may be determined by the use of a probability tree (see Worksheet 5, Question 15).

There are several standard types of discrete probability distribution. We will consider two of the most important, namely the binomial distribution and the Poisson distribution.

6.2 BINOMIAL DISTRIBUTION, AN EXAMPLE

Consider another coin-tossing example, but this time we will toss the coin 10 times. The number of heads will vary if we repeatedly toss the coin 10 times and we may note the following:

1. We have a fixed number of tosses, that is 10.
2. Each toss can result in one of only two outcomes, heads and tails.
3. The probability of heads is the same for all tosses, and is $\frac{1}{2}$ for a fair coin.
4. The tosses are independent in the sense that the probability of heads for any toss is unaffected by the result of previous tosses.

Because these four conditions are satisfied, the experiment of tossing the coin 10 times is an example of what is called a *binomial experiment* (consisting of 10 so-called 'Bernoulli' trials).

The variable 'number of heads in the 10 tosses of a coin' is said to have a *binomial distribution* with 'parameters' 10 and 0.5, which we write in a short-hand form as $B(10, 0.5)$. The first parameter, 10, is the number of trials or tosses and the second parameter, 0.5, is the probability of heads in a single trial or toss.

6.3 THE GENERAL BINOMIAL DISTRIBUTION

To generalize the example of the previous section, the outcomes of each trial in a binomial experiment are conventionally referred to as 'success' (one of the outcomes) and 'failure'. The general binomial distribution is denoted by $B(n, p)$ where the parameters n and p are the number of trials and the probability of success in a single trial, respectively.

In order to decide *a priori* whether a variable has a binomial distribution we must check the four conditions (generalizing on those of the previous section):

1. There must be a fixed number of trials, n.
2. Each trial can result in one of only two outcomes, which we refer to as success and failure.
3. The probability of success in a single trial, p, is constant.
4. The trials are independent, so that the probability of success in any trial is unaffected by the results of previous trials.

If all four conditions are satisfied then the discrete random variable, which we call x, to stand for the number of successes in n trials, has a $B(n, p)$ distribution.

Unless n is small ($\leqslant 3$, say) the methods we used in Chapter 5 are inefficient for calculating probabilities. Luckily we can use a formula, which we shall quote without proof, or in some cases we can use tables, see Section 6.6, to find probabilities for a particular binomial distribution, if we know the numerical values of the parameters, n and p.

The formula is

$$P(x) = \binom{n}{x} p^x (1 - p)^{n-x} \qquad \text{for } x = 0, 1, 2, \ldots, n.$$

This formula is not difficult to use if each part is understood separately: $P(x)$ means 'the probability of x successes in n trials'.

$\binom{n}{x}$ is a shorthand for $n!/x!(n - x)!$, where $n!$ means 'factorial n' (refer to Section 2.2 if necessary).

$x = 0, 1, 2, \ldots, n$ means that we can use this formula for each of these values of x, which are the possible numbers of successes in n trials.

6.4 CALCULATING BINOMIAL PROBABILITIES, AN EXAMPLE

For the example of tossing a coin 10 times, $n = 10$. If we assume the coin is fair and we regard 'heads' as the outcome we think of as a 'success', then $p = 0.5$. We can now calculate the probabilities of getting all possible numbers of heads using the formula:

$$P(x) = \binom{10}{x}(0.5)^x(1 - 0.5)^{10-x} \qquad \text{for } x = 0, 1, 2, \ldots, 10.$$

For example, to find the probability of getting 8 heads in 10 tosses we put $x = 8$ in this formula:

$$\begin{aligned} P(8) &= \binom{10}{8}(0.5)^8(1 - 0.5)^{10-8} \\ &= \frac{10!}{8!2!}(0.5)^8(0.5)^2 \\ &= 0.0439. \end{aligned}$$

Similarly for the other possible values of x, and recalling that $0! = 1$, we obtain Table 6.2.

Table 6.2 A binomial distribution for $n = 10, p = 0.5$

Number of heads, x	0	1	2	3	4	5	6	7	8	9	10
Probability, $P(x)$	0.001	0.010	0.044	0.117	0.205	0.246	0.205	0.117	0.044	0.010	0.001

Since this set of 11 possible events resulting from tossing a coin 10 times form a mutually exclusive and exhaustive set, the probabilities in Table 6.2 sum to 1 (recalling Section 5.11). The probabilities for any discrete probability distribution should sum to 1 in this way (apart from rounding errors, which may affect the last decimal place), and this fact provides a useful check on our calculations.

The information in Table 6.2 can also be presented graphically (see Fig. 6.1) in a graph which is similar to the line chart in Fig. 3.3, except that the vertical axis represents probability rather than frequency.

Notes

(a) The distribution is symmetrical about the centre-line at $x = 5$, but this is because $p = 0.5$ for this distribution, which is exactly in the middle of the possible range of 0 to 1 for a probability. Binomial distributions with p values greater than 0.5 will be negatively skew, those with p values less than 0.5 will be positively skew.

(b) Since the number of heads is a discrete random variable, the graph above is also discrete with gaps between the possible values.

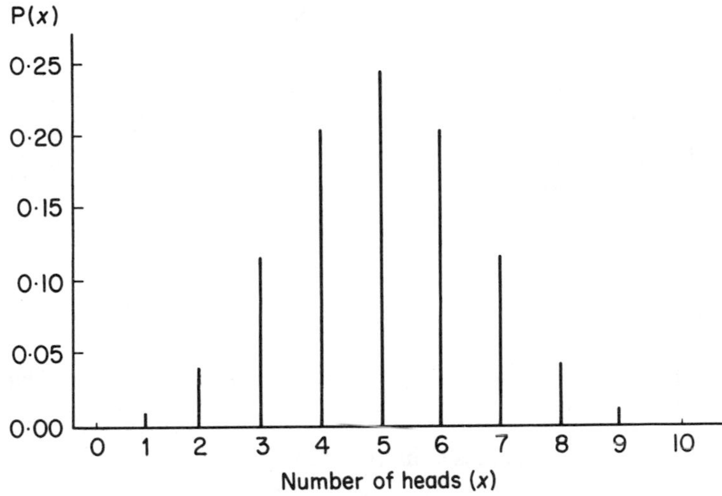

Fig. 6.1 A binomial distribution for $n = 10$, $p = 0.5$.

6.5 THE MEAN AND STANDARD DEVIATION OF THE BINOMIAL DISTRIBUTION

In Chapter 4 we calculated the mean and standard deviation of sample data. In a similar way we think of the mean and standard deviation of the binomial distribution, meaning the mean and standard deviation of the values taken by the variable in repeated binomial experiments. Instead of having to carry out these experiments in order to calculate the mean and standard deviation, it can be shown mathematically that the following formulae may be applied to any binomial distribution:

$$\text{mean} = np, \qquad \text{standard deviation} = \sqrt{(np(1-p))}.$$

Example

For the coin-tossing example of the previous section, $n = 10$, $p = 0.5$, so that the mean equals $10 \times 0.5 = 5$, and standard deviation equals $\sqrt{(10 \times 0.5 \times 0.5)} = 1.58$.

It should seem intuitively reasonable that the mean number of heads in repetitions of 10 tosses of a coin is 5. But what does a standard deviation of 1.58 tell us? As stated before, we shall derive more meaning from the value of a standard deviation when we discuss the normal distribution in Chapter 7. For the time being we remember that the larger the standard deviation, the more the variable will vary. So a variable with a $B(20, 0.5)$ distribution varies more than a variable with a $B(10, 0.5)$ distribution, since their standard deviations are 2.24 and 1.58 respectively.

6.6 USING TABLES TO OBTAIN BINOMIAL PROBABILITIES

To save time in calculating binomial probabilities Table D.1 of Appendix D may be used, instead of the formula for $P(x)$, for certain values of n and p. Table D.1 gives cumulative probabilities, that is the probabilities of 'so many or fewer successes'.

Example

$n = 10$, $p = 0.5$. In Table D.1, find the column of probabilities for these values of n and p.

To find $P(8) = P(8 \text{ successes in } 10 \text{ trials}, p = 0.5)$, find the row labelled $r = 8$, and read that:

$$P(8 \text{ or fewer successes in } 10 \text{ trials}, p = 0.5) = 0.9893.$$

Now find the row labelled $r = 7$ and read that:

$$P(7 \text{ or fewer successes in } 10 \text{ trials}, p = 0.5) = 0.9453.$$

Subtracting, $P(8 \text{ successes in } 10 \text{ trials}, p = 0.5) = 0.0440$. This agrees to 3 dps with the answer obtained in Section 6.4 using the formula.

In general, we use the idea that:

$$P(x \text{ successes}) = P(x \text{ or fewer successes}) - P((x - 1) \text{ or fewer successes}).$$

Example

$n = 10$, $p = 0.5$. Find $P(10)$.

$$P(10) = P(10 \text{ or fewer}) - P(9 \text{ or fewer}) = 1 - 0.9990 = 0.0010.$$

Because only certain values of n and p are given in the tables (since the tables would need to be very extensive to cover all possible values of n and p) you should learn how to use both the formula for $P(x)$ and Table D.1. The fact that no p values above 0.5 are given in Table D.1 is less of a restriction than it might first appear. This is because we can always treat as our 'success' the outcome which has the smaller probability. The following example illustrates this point and some other logical points.

Example

If 70% of male students eventually marry, what is the probability that 40 or more of a random sample of 50 male students will eventually marry? The outcome with the smaller probability is 'not marry', so we treat 'not marry' as 'success' and 'marry' as 'failure'. Assuming the four binomial conditions apply, use $n = 50$, $p = 1 - 0.7 = 0.3$.

$$\begin{aligned}\text{P(40 or more out of 50 will marry)} &= \text{P(40 or more `failures')}\\ &= \text{P(10 or fewer `successes')}\\ &= 0.0789,\end{aligned}$$

using Table D.1 for $n = 50$, $p = 0.3$, $r = 10$.

Note

In case you think this is a tortuous calculation, think of the alternative using the formula for P(x), this time thinking of 'marry' as 'success' and use $n = 50$, $p = 0.7$. Thus:

$$\begin{aligned}\text{P(40 or more out of 50 will marry)} &= \text{P}(40) + \text{P}(41) + \cdots + \text{P}(50)\\ &= \binom{50}{40}0.7^{40}(1 - 0.7)^{50-40}\\ &\quad + \text{ten similar terms}\end{aligned}$$

– a rather tedious calculation!

6.7 POISSON DISTRIBUTION, AN INTRODUCTION

The second standard type of discrete probability distribution which we will consider, the Poisson distribution, is concerned with the discrete variable number of random events per unit time or space. The words 'random' in this context implies that there is a constant probability that the event will occur in one unit of time or space (space can be one-, two- or three-dimensional).

6.8 SOME EXAMPLES OF POISSON VARIABLES

There are many examples which may be used to illustrate the great variety of applications of the Poisson distribution as a 'model' for random events:

(a) At the telephone switchboard in a large office block there may be a constant probability that a telephone call will be received in a given minute. The number of calls received per minute will have a Poisson distribution.
(b) In spinning wool into a ball from the raw state, there may be a constant probability that the spinner will have to stop to untangle a knot. The number of stops per 100 metres of finished wool will then have a Poisson distribution.
(c) In the production of polythene sheeting there may be a constant probability of a blemish (called 'fish-eyes') which makes the film unsightly or opaque. The number of blemishes per square metre will then have a Poisson distribution.

Other examples concerned with random events in time are the number of postilions killed per year by lightning in the days of horse-drawn carriages, the number of major earthquakes recorded per year and the number of α-particles emitted per unit time by a radioactive source.

6.9 THE GENERAL POISSON DISTRIBUTION

In order to decide whether a discrete random variable has a Poisson distribution we must be able to answer 'Yes' to the following questions:

1. Are we interested in random events per unit time or space?
2. Is the number of events which might occur in a given unit of time or space theoretically unlimited?

If the answer to the first question is yes but the answer to the second question is no, the distribution may be binomial – check the four conditions in Section 6.3.

In order to calculate the probabilities for a Poisson distribution we can use a formula, which we shall quote without proof, or in some cases we can use tables (see Section 6.12) if we know the numerical value of the parameter m (defined below) for a particular Poisson distribution.

The formula is

$$\mathrm{P}(x) = \frac{\mathrm{e}^{-m}m^x}{x!} \qquad \text{for } x = 0, 1, 2, \ldots$$

$\mathrm{P}(x)$ means the probability of x random events per unit time or space, e is the number 2.718...(refer to Section 2.4 if necessary), m is the mean number of events per unit time or space, $x = 0, 1, 2, \ldots$ means we can use this formula for 0 and any positive whole number.

6.10 CALCULATING POISSON PROBABILITIES, AN EXAMPLE

Example

Telephone calls arrive randomly at a switchboard, one a minute on average. What are the probabilities that 0, 1, 2,... calls will be received in a period of 5 minutes?

Since the probabilities of interest relate to a unit of time of 5 minutes we calculate m as the mean number of calls per 5 minutes. So $m = 5$ for this example, and

$$\mathrm{P}(x) = \frac{\mathrm{e}^{-5}5^x}{x!} \qquad \text{for } x = 0, 1, 2, \ldots$$

For example, for $x = 6$ calls in 5 minutes,

$$P(6) = \frac{e^{-5}5^6}{6!} = \frac{0.00674 \times 15\,625}{720} = 0.146.$$

Similarly for other values of x, and recalling that $0! = 1$, we obtain Table 6.3.

Table 6.3 A Poisson distribution for $m = 5$

Number of calls received in 5 minutes (x)	0	1	2	3	4	5	6	7	8	9	10
Probability, P(x)	0.007	0.034	0.084	0.140	0.176	0.176	0.146	0.104	0.065	0.036	0.018

These 11 possible events (in observing a switchboard for 5 minutes) form a mutually exclusive set but they are not an exhaustive set since x values above 10 are possible although their probabilities are small. The probabilities in Table 6.3 sum to 0.986, which illustrates the same point.

The information in Table 6.3 is also presented graphically in Fig. 6.2 (in a graph similar to Fig. 6.1 for a binomial distribution).

Note that the distribution is slightly positively skewed. Poisson distributions with values of m less than 5 will be more positively skewed;

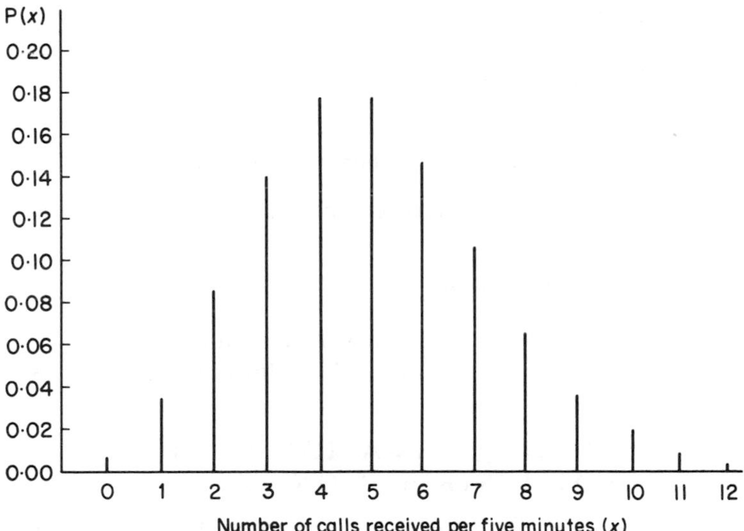

Fig. 6.2 A Poisson distribution for $m = 5$.

those with values of m greater than 5 will be less skewed and hence more symmetrical.

6.11 THE MEAN AND STANDARD DEVIATION OF THE POISSON DISTRIBUTION

The mean of the Poisson distribution is m, as already stated in Section 6.9, and the standard deviation of the Poisson distribution is $\sqrt{m}$.

For the example above, the mean is 5 and the standard deviation is 2.24.

6.12 USING TABLES TO OBTAIN POISSON PROBABILITIES

To save time in calculating Poisson probabilities, Table D.2 of Appendix D may be used for certain values of m, instead of the formula for $P(x)$. In a similar way to Table D.1 of Appendix D, Table D.2 gives cumulative probabilities, that is the probabilities of 'so many or fewer random events per unit time or space'.

Example

$m = 5$. In Table D.2 find the column of probabilities for this value of m.

To find $P(6) = P(6$ random events when $m = 5)$, find the row labelled $r = 6$, and read that:

$$P(6 \text{ or fewer random events when } m = 5) = 0.7622.$$

Now find the row labelled $r = 5$ and read that:

$$P(5 \text{ or fewer random events when } m = 5) = 0.6160.$$

Subtracting,

$$P(6 \text{ random events when } m = 5) = 0.1462.$$

This agrees to 3 dps with the answer obtained in Section 6.10 using the formula.

In general we use the idea that:

$$P(x \text{ random events}) = P(x \text{ or fewer random events}) - P((x-1) \text{ or fewer random events}).$$

*6.13 POISSON APPROXIMATION TO THE BINOMIAL DISTRIBUTION

There are examples of binomial distributions for which the calculation of approximate probabilities is made easier by the use of the formula or tables for the Poisson distribution! Such an approach can be justified theoretically for binomial distributions with large values of n and small values of p. The resulting probabilities are only approximate, but quite good approximations may be obtained when $p < 0.1$, even if n is not large, by putting $m = np$.

Example

Assume 1% of people are colour-blind. What is the probability that 10 or more randomly chosen from 500 will be colour-blind?

This is a 'binomial' problem with $n = 500$ and $p = 0.01$. But Table D.1 cannot be used for $n = 500$, and to use the binomial formula we would need to calculate $1 - P(0) - P(1) - \cdots - P(9)$.

However, using the Poisson approximation for $m = np = 5$, and Table D.2 for $r = 9$, we read that

$$P(9 \text{ or fewer colour-blind in } 500) = 0.9682.$$

Hence, $P(10 \text{ or more colour-blind in } 500) = 1 - 0.9682 = 0.0318$.

6.14 SUMMARY

The binomial and the Poisson distributions are two of the most important discrete probability distributions.

The binomial distribution gives the probabilities for the numbers of successes in a number of trials, if four conditions hold. Binomial probabilities may be calculated using:

$$P(x) = \binom{n}{x} p^x (1 - p)^{n-x}$$

or, in certain cases, Table D.1.

The Poisson distribution gives the probabilities for the number of random events per unit time or space. Poisson probabilities may be calculated using:

$$P(x) = \frac{e^{-m} m^x}{x!}$$

or, in certain cases, Table D.2.

If $p < 0.1$, it may be preferable to calculate binomial probabilities using the Poisson approximation.

WORKSHEET 6: THE BINOMIAL AND POISSON DISTRIBUTIONS

Questions 1–12 are on the binomial distribution.

1. For the binomial distribution, what do n and p stand for?

2. How can you tell *a priori* whether a discrete random variable has a binomial distribution?

3. In a binomial experiment each trial can result in one of two outcomes. Which one shall I call a 'success' and which 'failure'?

4. What is the general name for the variable which has a binomial distribution?

5. For the distribution B (3, 0.5),
   (a) How many outcomes are there to each trial?
   (b) How many trials are there?
   (c) How many possible values can the variable take?
   (d) What are the mean and standard deviations of this distribution?
   (e) Is it a symmetrical distribution?

6. For families with four children, what are the probabilities that a randomly selected family will have 0, 1, 2, 3, 4 boys, assuming that boys and girls are equally likely at each birth? Check that the probabilities sum to 1. Why do they?

   Given 200 families each with four children, how many would you expect to have 0, 1, 2, 3, 4 boys?

7. In a multiple choice test there are five possible answers to each of 20 questions. If a candidate guesses the answer each time:
   (a) What is the mean number of correct answers you would expect the candidate to obtain?
   (b) What is the probability that the candidate will pass the test by getting 8 or more correct answers?
   (c) What is the probability that the candidate will get at least one correct answer?

8. 5% of items in a large batch are defective. If 50 items are selected at random, what is the probability that:
   (a) At least one will be defective?
   (b) Exactly two will be defective?
   (c) Ten or more will be defective?

   Use tables to answer these questions initially, but check the answers to (a) and (b) using a formula.

9. In an experiment with rats, each rat goes into a T-maze in which there is a series of T-junctions. At each junction a rat can either turn left or right. Assuming a rat chooses at random, what are the probabilities that it will make 0, 1, 2, 3, 4 or 5 right turns out of 5 junctions?

10. A new method of treating a disease has a 70% chance of resulting in a cure. Show that if a random sample of 10 patients suffering from the disease are treated by this method the chance that there will be more

than 7 cures is about 0.38. What other word could be used in place of 'chance'?

11. 50 g of yellow wallflower seeds are thoroughly mixed with 200 g of red wallflower seeds. The seeds are then bedded out in rows of 20.
    (a) Assuming 100% germination, why should the number of yellow wallflower plants per row have a binomial distribution?
    (b) What are the values of n and p for this distribution?
    (c) What is the probability of getting a row with:
        (i) No yellow wallflower plants in it?
        (ii) One or more yellow wallflower plants in it?

12. A supermarket stocks eggs in boxes of six, and 10% of the eggs are cracked. Assuming that the cracked eggs are distributed at random, what is the probability that a customer will find that the first box he chooses contains:
    (a) No cracked eggs?
    (b) At least one cracked egg?

    If he examines five boxes what is the probability that three or more will contain no cracked eggs?

Questions 13–19 are on the Poisson distribution.

13. For the Poisson distribution we use the formula

$$\mathrm{P}(x) = \frac{\mathrm{e}^{-m} m^x}{x!}.$$

    What do the symbols m, e, x stand for? What values can x take?

14. If a variable has a Poisson distribution with a mean of 4, what are its standard deviation and variance? What can you say about the values of the mean and variance of any Poisson distribution?

15. The Poisson distribution is the distribution of the number of random events per unit time. What does the word 'random' mean here?

16. Assuming that breakdowns in a certain electricity supply occur randomly with a mean of one breakdown every 10 weeks, calculate the probabilities of 0, 1, 2 breakdowns in any period of one week.

17. Assume that the number of misprints per page in a certain type of book has a Poisson distribution with a mean of one misprint per five pages. What percentage of pages contains no misprints? How many pages would you expect to have no misprints in a 500-page book?

18. A hire firm has three ladders which it hires out by the day. Records show that the mean demand over a period is 2.5 ladders per day. If it is assumed that the demand for ladders follows a Poisson distribution, find:

(a) The percentage of days on which no ladder is hired.
(b) The percentage of days on which all three ladders are hired.
(c) The percentage of days on which demand outstrips supply.

19. A roll of cloth contains an average of 3 defects per 100 square metres distributed at random. What is the probability that a randomly chosen section of 100 square metres of cloth contains:
(a) No defects?
(b) Exactly 3 defects?
(c) 3 or more defects?

Questions 20–22 are on the Poisson approximation to the binomial distribution.

*20. A rare blood group occurs in only 1% of the population, distributed at random. What is the probability that at least one person in a random sample of 100 has blood of this group? Use both the 'binomial' method, and the 'Poisson approximation to the binomial' method. Compare your answers. Which is correct?

*21. If, in a given country, an average of one miner in 2000 loses his life due to accident per year, calculate the probability that a mine in which there are 8000 miners will be free from accidents in a given year.

*22. The average number of defectives in batches of 50 items is 5. Find the probability that a batch will contain:
(a) 10 or more defectives.
(b) Exactly 5 defectives.

Use both the binomial and the Poisson approximation to the binomial methods, and compare your answers.

Continuous probability distributions

7.1 INTRODUCTION

In Chapter 3 we considered an example of a continuous variable, namely the height of a student, and we summarized the heights of 50 students by a histogram, Fig. 3.1, reproduced here as Fig. 7.1.

We saw in Worksheet 5, Question 6, how to express probability as the ratio of two areas, so that we could make statements such as

$$\text{P(randomly selected student has a height between 175 and 185 cm)} = \frac{\text{Area of rectangle on base 175–185}}{\text{Total area of histogram}}.$$

Suppose we apply this idea to the heights of all first-year students in higher education. In the histogram the number of students in each group would be much greater so we could have many more than five groups, and still have a fairly large number of students in each group. This histogram would look something like Fig. 7.2, where the vertical sides of the rectangles have been omitted and the top of the rectangles have been smoothed into a curve.

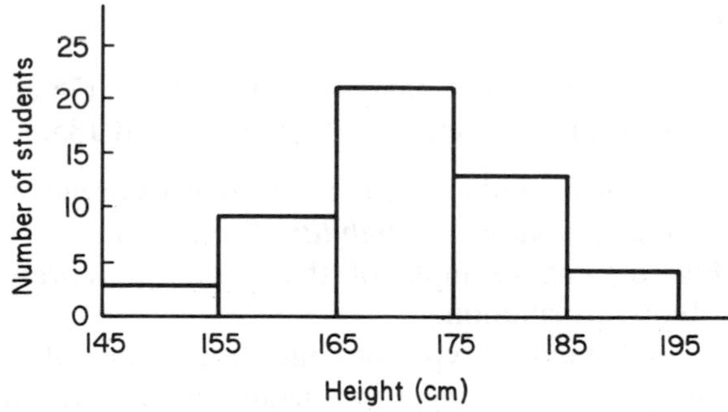

Fig. 7.1 Histogram of the heights of 50 students.

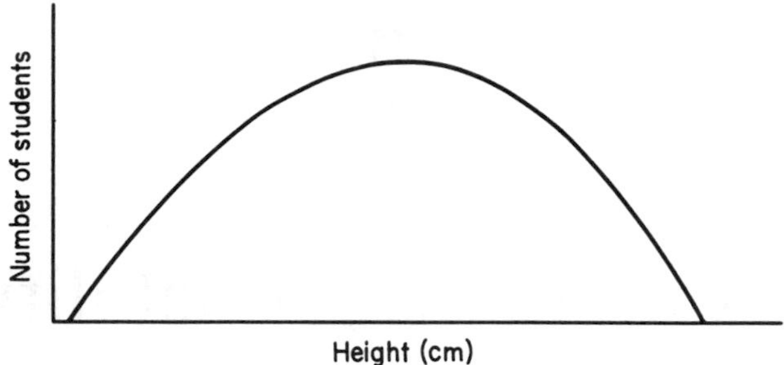

Fig. 7.2 Histogram of the heights of all first year students in higher education.

If this graph is 'scaled' in the vertical direction so that the total area under the curve was 1 in some units, then we would be wrong to keep calling the vertical axis 'Number of students'. But this curve would have the property that the probability of a height between any two values would be equal to the area under the curve between these values (Fig. 7.3).

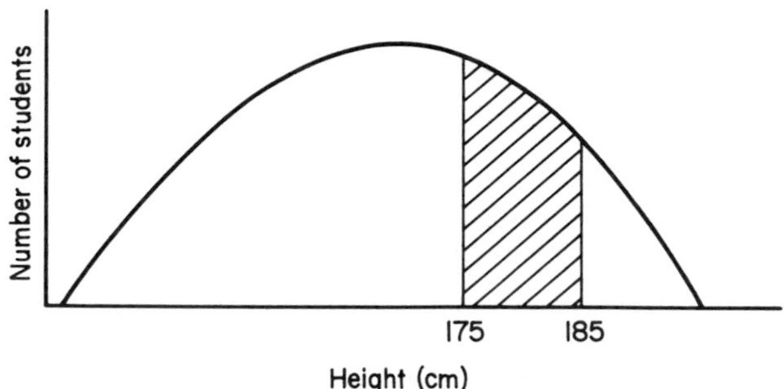

Fig. 7.3 Continuous probability distribution for the variable height.

For example,

P(randomly selected student has a height between 175 and 185 cm)
= Area under curve between 175 and 185.

Assuming such a curve could be drawn, this is an example of the graphical representation of a *continuous probability distribution.*

Compare Fig. 6.1, an example of the graphical representation of a discrete probability distribution.

There are several standard types of continuous probability distribution. We will consider two of the most important, namely the normal distribution and the rectangular (or uniform) distribution.

7.2 THE NORMAL DISTRIBUTION

The normal distribution is the most important distribution in statistics. There are two main reasons for this:

1. It arises when a variable is measured for a large number of nominally identical objects, and when the variation may be assumed to be caused by a number of factors each exerting a small positive or negative random influence on an individual object. An example is the variable 'height of students', where the variation in heights is caused by many factors such as age, sex, diet, exercise, height of parents, bone structure, etc.
2. The properties of the normal distribution have a very important use in the statistical theory of drawing conclusions from sample data about the populations from which the samples are drawn (these methods will be discussed in Chapter 8 onwards).

Returning to the idea of graphically representing distributions, the normal distribution has a bell-shape (with most values concentrated in the centre, and few extreme values), is unimodal, and is symmetrical (Fig. 7.4). It has two parameters μ and σ.†

There are a number of related properties of the normal distribution which (at last!) give us a better understanding of the meaning of standard deviation as a measure of variation.

1. Approximately 68% of the area of a normal distribution lies within one standard deviation of the mean. So the area between the vertical lines drawn at $\mu - \sigma$ and $\mu + \sigma$ in Fig. 7.4 is roughly two-thirds of the total area.

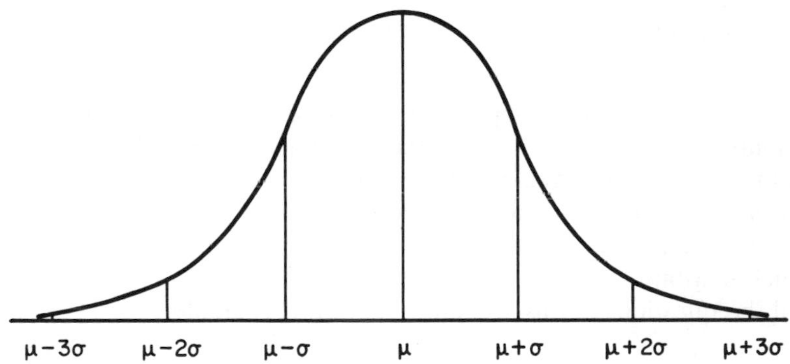

Fig. 7.4 The normal distribution: μ is the mean of the distribution, σ is the standard deviation of the distribution.

† *Important note on notation* In Chapter 4 the symbols $\bar{x}$ and s were used to denote the *sample* mean and *sample* standard deviation, respectively. The Greek letters μ (lower case mu) and σ (low case sigma) are used here because we are now dealing with a *population* of measurements. Samples and populations are defined and discussed more fully in Chapter 8.

2. Approximately 95% of the area of a normal distribution lies within two standard deviations of the mean (exactly 95% of the area lies within 1.96 standard deviations of the mean).
3. Approximately 99.7% of the area of a normal distribution lies within three standard deviations of the mean.

7.3 AN EXAMPLE OF A NORMAL DISTRIBUTION

Suppose we know that the variable height (of all first-year students in higher education) is normally distributed with a mean $\mu = 170$ cm and standard deviation $\sigma = 10$ cm.

Using the properties stated in the previous section we could state that:

1. Approximately 68% have heights between $170 - 10 = 160$ and $170 + 10 = 180$ cm.
2. Approximately 95% have heights between $170 - (2 \times 10) = 150$ and $170 + (2 \times 10) = 190$ cm.
3. Approximately 99.7% have heights between $170 - (3 \times 10) = 140$ and $170 + (3 \times 10) = 200$ cm.

The first statement is equivalent to 'the probability that a randomly selected student will have a height between 160 and 180 cm is 0.68'.

But how can we calculate probabilities and percentages for other heights of interest? The answer is that we need to be able to calculate the areas under the normal distribution curve, and we do this using Table D.3(a) of Appendix D.

> Table D.3(a) enables us to calculate probabilities for any normal distribution if we know the numerical values of μ and σ for that distribution.

Table D.3(a) actually gives probabilities in terms of areas of the normal distribution curve; that is areas to the left of particular values.

Let us consider the example of the normal distribution of heights with a mean $\mu = 170$ and standard deviation $\sigma = 10$. (In shorthand form this would be referred to as the $N(170, 10^2)$ distribution, where the general normal distribution is $N(\mu, \sigma^2)$.)

All the questions in this section refer to this example.

Question 7.1

What is the probability that a randomly selected first-year student in higher education will have a height greater than 185 cm?

The answer is the area to the right of 185 in Fig. 7.5, since 'to the right of 185' implies 'greater than 185'.

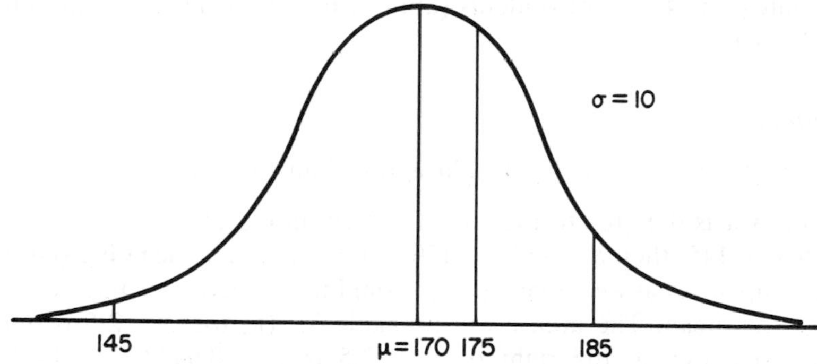

Fig. 7.5 A normal distribution with $\mu = 170$, $\sigma = 10$.

In order to use Table D.3(a) we first have to 'transform' our normal distribution into one with a mean $\mu = 0$ and standard deviation $\sigma = 1$ (the so-called standardized normal distribution). We do this by calculating z values using the formula $z = (x - \mu)/\sigma$. Let us see how to apply this to our example:

Since we are interested in the value 185, let $x = 185$.

Now calculate the z value, using $z = (x - \mu)/\sigma$. So

$$z = \frac{185 - 170}{10} = 1.5.$$

Using Table D.3(a) for $z = 1.5$ we read that the area to the left of 185 is 0.9332. Hence the area to the right of 185 is $1 - 0.9332 = 0.0668$. We can also state that 6.68% of students (about 1 in 15) have a height greater than 185 cm (by multiplying the probability by 100).

Question 7.2

What is the probability that height lies between 175 and 185 cm?

The answer is the area between 175 and 185 in Fig. 7.5, which we can think of as

Area to the left of 185 − Area to the left of 175.

Since we are now interested in the value 175, let $x = 175$.

Now calculate $z = (x - \mu)/\sigma = (175 - 170)/10 = 0.5$. Using Table D.3(a) for $z = 0.5$, we read that the area to the left of 175 is 0.6915. Therefore the area between 175 and 185 $= 0.9332 - 0.6915 = 0.2417$, using the answer to Question 7.1.

The probability that height lies between 175 and 185 is 0.2417. We can also state that 24.17% of students (about 1 in 4) have a height between 175 and 185 cm.

Question 7.3

What is the probability that height is less than 145 cm?

The answer is the area to the left of 145 cm in Fig. 7.5.

Let $x = 145$, then $z = (145 - 170)/10 = -2.5$. The negative sign for z simpy implies what we already know from Fig. 7.5 that the value of 145 lies below the mean. The area given in Table D.3(a) for $z = 2.5$ is 0.9938. Hence the area to the right of $z = 2.5$ is $1 - 0.9938 = 0.0062$. By symmetry, this is also the area to the left of $z = -2.5$. So the required probability is 0.0062.

Question 7.4

What is the probability that height lies between 145 and 175 cm?

From previous answers, the required probability is:

$$0.6915 - 0.0062 = 0.6853$$

Question 7.5

What is the probability that height is less than 170 cm?

By the symmetry of the normal distribution of Fig. 7.5, the answer is 0.5, or we could use $x = 170$, $z = (170 - 170)/10 = 0$, and Table D.3(a).

7.4 COMPARISON OF DIFFERENT NORMAL DISTRIBUTIONS

There is not just one normal distribution but a 'family' depending on the particular values of μ and σ.

Suppose we consider the three normal distributions with parameters

1. $\mu = 100$, $\sigma = 5$
2. $\mu = 120$, $\sigma = 5$
3. $\mu = 100$, $\sigma = 10$

Using the properties that the total area of each must be 1, and that nearly all (95%) of the area lies with two standard deviations of the mean, Fig. 7.6 shows a rough comparison between the distributions.

To get from distribution (1) to (2) we shift the curve 20 units to the right, keeping exactly the same shape. To get from distribution (1) to (3) we keep

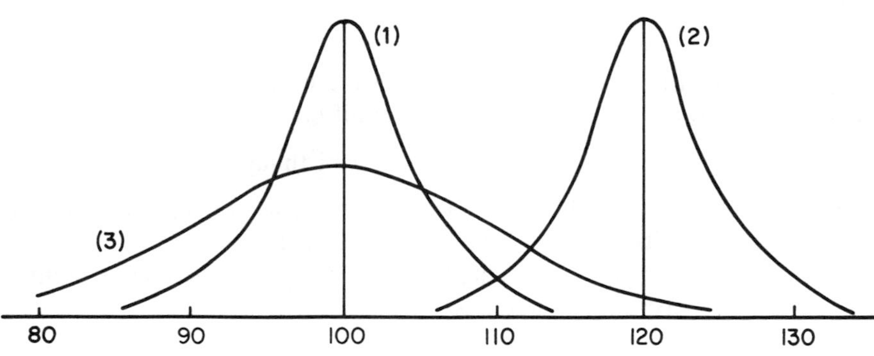

Fig. 7.6 Comparison of three normal distributions.

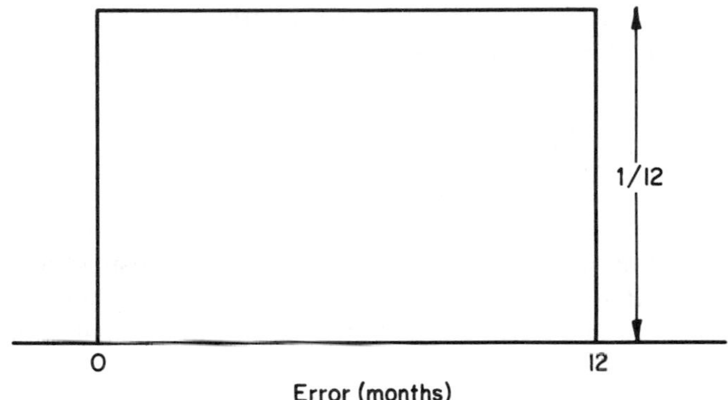

Fig. 7.7 A rectangular distribution for 'error' in stated age.

the mean value at 100, but depress the curve and spread it out more, keeping the same total area.

7.5 RECTANGULAR DISTRIBUTION

Because of the great importance of the normal distribution, students who take introductory courses in statistics often believe that all continuous variables are normally distributed. Partly to counteract this erroneous belief at any early stage (it will also be counteracted in Chapter 11 when we deal with inferential methods which deal specifically with non-normal continuous variables) we now introduce another continuous probability distribution, the rectangular (also called the uniform) distribution.

This is a rather dull and flat distribution, see Fig. 7.7, but it does have the advantage that probabilities are easy to calculate.

Example

Suppose we consider the 'error' which is made when a person states his or her 'age at last birthday'. The error is the difference

actual age − age at last birthday

and this continuous variable is equally likely to lie anywhere in the range 0 to 12 months, so its probability distribution is as in Fig. 7.7.

Since the total area must be equal to 1, the height of the rectangle must be equal to 1/(base of rectangle) = 1/12.

Question 7.6

What percentage of errors will be less than 3 months?

The probability of an error of less than 3 months is the area to the left of 3 which is $3 \times 1/12 = 1/4$, so 25% of errors will be less than 3 months.

*7.6 THE NORMAL APPROXIMATION TO THE BINOMIAL DISTRIBUTION

Just as there are conditions (see Section 6.13) when the calculation of approximate binomial probabilities is made easier by using the formula or tables for the Poisson distribution, there are other conditions when it is preferable to use the normal distribution tables to calculate approximate binomial probabilities.

We may use this 'normal approximation to the binomial distribution' when $np > 5$, $n(1 - p) > 5$, and when Table D.1 cannot be used. The conditions on n and p are more likely to be met when n is large and p is not extreme.

Example

Suppose one person in six is left-handed. If a class contains 40 students, what is the probability that 10 or more are left-handed? Assuming the four conditions for the binomial apply this is a 'binomial' problem with $n = 40$, $p = 1/6$, so $np = 6.67$ and $n(1 - p) = 33.33$, and hence the conditions for using the normal approximation to the binomial distribution are satisfied. We may therefore treat the variable 'number of left-handed players in a sample of 40' as though it was normally distributed with:

$$\mu = np = 6.67, \qquad \sigma = \sqrt{(np(1 - p))} = 2.36.$$

This distribution is shown in Fig. 7.8.

Before we use Table D.3(a) we must apply a 'continuity correction' of $\frac{1}{2}$ since the 'number of left-handed players' is a discrete variable while the

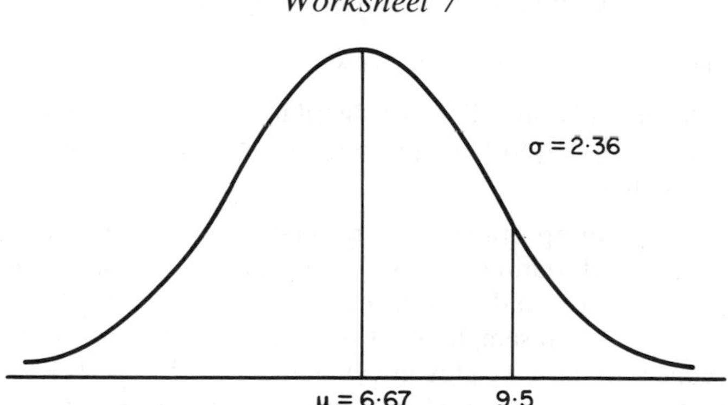

Fig. 7.8 A normal distribution with $\mu = 6.67$, $\sigma = 2.36$.

normal distribution is a continuous probability distribution. Since '10 or more on a discrete scale' is equivalent to 'more than 9.5 on a continuous scale', we use Table D.3(a) for $x = 9.5$. So:

$$z = \frac{x - \mu}{\sigma} = \frac{9.5 - 6.67}{2.36} = 1.2,$$

giving an area to the left of 9.5 of 0.8849, using Table D.3(a).

The probability of 10 or more left-handed students in 40 is $1 - 0.8849 = 0.1151$.

7.7 SUMMARY

The normal and the rectangular distributions are two standard types of continuous probability distributions. The normal distribution is the most important in statistics because it arises when a number of factors exert small positive or negative effects on the value of the variable, and because it is extremely useful in the theory of statistical inference.

Probabilities and percentages for the normal distribution are calculated using tables if we have values of μ and σ, and for the rectangular distribution by calculating the areas of rectangles. The total area under any continuous distribution curve is 1.

The normal distribution tables may also be used to calculate approximate binomial probabilities if $np > 5$ and $n(1 - p) > 5$, and where the use of the (binomial) Table D.1 is not possible.

WORKSHEET 7: THE NORMAL AND RECTANGULAR DISTRIBUTIONS

1. For the normal distribution, what do μ and σ stand for?

2. Is a normal distribution ever skew?

3. For the binomial and Poisson distributions the probabilities sum to 1. What is the equivalent property for the normal and rectangular distributions?

4. The weights of 5p oranges are normally distributed with mean 70 g and standard deviation 3 g. What percentage of these oranges weigh: (a) over 75 g, (b) under 60 g, (c) between 60 and 75 g?

   Given a random sample of 50 5p oranges, how many of them would you expect to have weights in categories (a), (b) and (c)?

   Suppose the oranges are kept for a week and each orange loses 5 g in weight, re-work this question.

5. A fruit grower grades and prices oranges according to their diameter:

Diameter (cm)	*Price per orange* (p)
Below 5	4
Above 5 and below 6	5
Above 6 and below 7	6
Above 7 and below 8	7
Above 8 and below 9	8
Above 9 and below 10	9
Above 10	10

   Assuming that the diameter is normally distributed with a mean of 7.5 cm and standard deviation 1 cm, what percentage of oranges will be priced at each of the above prices? Given 10 000 oranges: (a) what is their total price, (b) what is their mean price?

6. A machine produces components whose thicknesses are normally distributed with a mean of 0.4 cm and standard deviation 0.01 cm. Components are rejected if they have a thickness outside the range 0.38 cm to 0.41 cm. What percentage are rejected? Show that the percentage rejected will be reduced to its smallest value if the mean is reduced to 0.395 cm, assuming the standard deviation remains unchanged.

7. Guests at a large hotel stay for an average of 9 days with a standard deviation of 2.4 days. Among 1000 guests how many can be expected to stay: (a) less than 7 days, (b) more than 14 days, (c) between 7 and 14 days? Assume that length of stay is normally distributed.

8. It has been found that the annual rainfall in a town has a normal distribution, because of the varying pattern of depressions and

anticyclones. If the mean annual rainfall is 65 cm, and in 15% of years the rainfall is more than 85 cm, what is the standard deviation of annual rainfall? What percentage of years will have a rainfall below 50 cm?

9. In the catering industry the wages of a certain grade of kitchen staff are normally distributed with a standard deviation of £4. If 20% of staff earn less than £30 a week, what is the average wage? What percentage of staff earn more than £50 a week?

10. Sandstone specimens contain varying percentages of void space. The mean percentage is 15%, the standard deviation is 3%. Assuming the percentage of void space is normally distributed, what proportion of specimens have a percentage of void space: (a) below 15%, (b) below 20%, (c) below 25%?

11. The height of adult males is normally distributed with a mean of 172 cm and standard deviation 8 cm. If 99% of adult males exceed a certain height, what is this height?

12. A machine which automatically packs potatoes into bags is known to operate with a mean of 25 kg and a standard deviation of 0.5 kg. Assuming a normal distribution what percentage of bags weigh: (a) more than 25 kg, (b) between 24 and 26 kg?

    To what new target mean weight should the machine be set so that 95% of bags weigh more than 25 kg? In this case what weight would be exceeded by 0.1% of bags?

*13. Using tables, show that for any normal distribution with mean μ and standard deviation σ:
(a) 68.26% of the area lies between $(\mu - \sigma)$ and $(\mu + \sigma)$.
(b) 95.45% of the area lies between $(\mu - 2\sigma)$ and $(\mu + 2\sigma)$.
(c) 99.73% of the area lies between $(\mu - 3\sigma)$ and $(\mu + 3\sigma)$.
(d) 5% of the area lies outside the interval from $(\mu - 1.96\sigma)$ to $(\mu + 1.96\sigma)$.
(e) 5% of the area lies above $(\mu + 1.645\sigma)$.

*14. A haulage firm has 60 lorries. The probability that a lorry is available for business on any given day is 0.8. Find the probability that on a given day:
(a) 50 or more lorries are available.
(b) Exactly 50 lorries are available.
(c) Less than 50 lorries are available.

15. An office worker commuting to London from Oxford each day keeps a record of how late his train arrives in London. He concludes that the train is equally likely to arrive anywhere between 10 minutes early and 20 minutes late. If the train is more than 15 minutes late the office worker will be late for work. How often will this happen?

Repeat this question assuming now that the number of minutes late is normally distributed with $\mu = 5$ minutes, $\sigma = 7.5$ minutes. Represent both distributions in a sketch indicating the answers as areas in the sketch.

Samples and populations

> We should extend our views far beyond the narrow bounds of a parish, we should include large groups of mankind.

8.1 INTRODUCTION

In the remaining chapters we shall mainly be concerned with statistical inference which will cover methods of drawing conclusions from sample data about the larger populations from which the samples are drawn. Although we have met the words 'sample' and 'population' in earlier chapters they were not defined. Here are the definitions:

Population

A population is the whole set of measurement about which we want to draw a conclusion. If we are interested in only one variable we call the population 'univariate'; for example, the heights of all first-year students in higher education form a univariate population. Notice that a population is the set of measurements, not the individuals or objects on which the measurements are made.

Sample

A sample is a sub-set of a population, a set of some of the measurements which comprise the population.

8.2 REASONS FOR SAMPLING

The first and obvious reason for sampling is to save time, money and effort. For example, in opinion polls before a General Election it would not be practicable to ask the opinion of the whole electorate on how it intends to vote.

The second and less obvious reason for sampling is that, even though we have only part of all the information about the population, nevertheless the sample data can be useful in drawing conclusions about the population, provided that we use an appropriate sampling method (see Section 8.3) and choose an appropriate sample size (see Section 8.4).

The third reason for sampling applies to the special case where the act of measuring the variable destroys the 'individual', such as in the destructive testing of explosives. Clearly testing a whole batch of explosives would be ridiculous.

8.3 SAMPLING METHODS

There are many ways of selecting a sample form a population but the most important is *random sampling* and the methods of statistical inference used in this book apply only to cases in which the sampling method is effectively random. Nevertheless there are other sampling methods which are worth mentioning for reasons stated below.

A *random sample* is defined as one for which each measurement in the population has the same chance (probability) of being selected. A sample selected so that the probabilities are not the same for each measurement is said to be a *biased sample*. For the population of student heights, selecting only students who were members of male basket-ball teams might result in a biased sample since nearly all these students would be above the mean height of all students. Random sampling requires that we can identify all the individuals or objects which provide the population, and that each measurement to be included in the sample is chosen using some method of ensuring equal probability, for example by the use of random number tables, such as Table D.4 of Appendix D.

Example

For the student height example, suppose that each student is assigned a unique five-digit number.[†] Starting anywhere in Table D.4 read off 5 consecutive digits by moving from one digit to the next in any direction – up, down, left, right or diagonally. This procedure could result in the number 61978, say. This number identifies the first student and his/her height is the first sample value. If there is no student with this number we ignore it and select another random number. We carry on until the required number of sample values, called the sample size, has been reached.

[†] Assuming that there are fewer than 100 000 students in total.

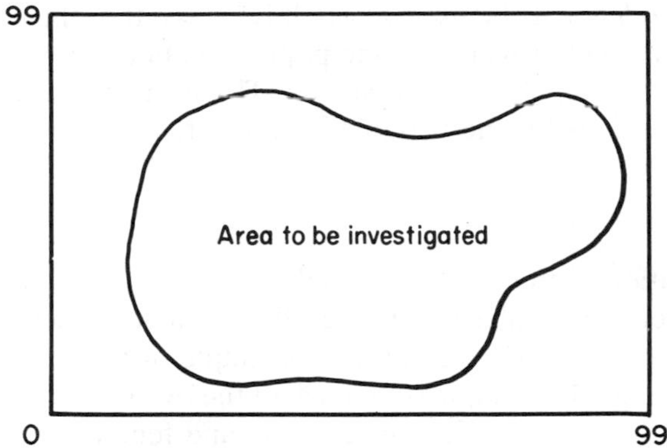

Fig. 8.1 Random sampling from a two-dimensional area.

Random numbers are also available on most scientific calculators and on computers.

Where the individuals are objects which are fixed in a given location, the method of random sampling can also be used. There are many examples in geography, geology and environmental biology where such samples are required. One method of obtaining a random sample is to overlay a plan of the area to be investigated with a rectangular grid which includes all points of interest in the area (see Fig. 8.1).

From one corner of the rectangle one side is labelled with the 100 values, 00 to 99, spaced at equal intervals. The other side is labelled in the same way. By selecting two 2-digit random numbers, the co-ordinates of a randomly selected point within the rectangle are determined. If this point falls outside the area of interest it is ignored and two more random numbers are selected. The procedure is repeated until the required number of points has been selected.

Systematic sampling may be used to cut down the time taken in selecting a random sample. In the student height example suppose we require a 1% (1 in 100) sample from 80 000 students. We randomly select one value in the range 1 to 100 and derive all other student numbers to be included in the sample by adding 100, 200, 300, and so on. So if the first number selected is 67, the others will be 167, 267, 367, . . . (giving a sample size of 800 in all).

Systematic sampling is suitable in that it provides a quasi-random sample so long as there is no periodicity in the population list or geographical arrangement which coincides with the selected values. For example, if we were selecting from a population of houses in an American city where the streets are laid out in a grid formation, then selecting every 100th house could result in always choosing a corner house on a cross-roads, possibly giving a biased sample.

Stratified sampling may be used where it is known that the individuals or objects to be sampled provide not one population but a number of distinct homogeneous sub-populations or strata. These strata may have quite different distributions for the variable of interest.

Example

For the population of the heights of students we may reasonably suppose that male students are taller on average than female students. If 60% of students are male and 40% female, a 1% sample of all students could be obtained by taking a 1% sample from each of the two strata. (We could test our hypothesis about the difference in male and female heights once we have obtained the sample data, using the methods of Chapter 10.)

Other methods of sampling include quota sampling, cluster sampling, multi-stage sampling and sequential sampling. Each method has its own special application area and will not be discussed here. Information on these methods may be found in specialized texts (see Appendix E).

8.4 SAMPLE SIZE

The most common question which every investigator who wishes to collect and analyse data asks is, 'How much data should I collect?' We refer to the number of values of a variable to be included in the sample as the *sample size* so the investigator is probably asking, 'What sample size should I choose?' If we wish to give a slightly more sophisticated answer than, for example, 'a sample size of 20 seems a bit too small, but 30 sounds about right', we can use the arguments of the following example.

Example

Suppose we wish to estimate the (population) mean height, μ, of students. The required sample size should depend on two factors:

1. The precision we require for the estimate, which the investigator must specify, knowing that the more precision he requires, the larger and sample size must be.
2. The variability of height, as measured by its standard deviation. We would think it reasonable to believe that the larger the standard deviation of height, the larger will be the required sample size. The standard deviation can only be determined when we have some data. But since we are trying to decide how much data to collect, we are in a chicken-and-egg situation.

We will return to a discussion of sample size when we have discussed the ideas of confidence intervals in the next chapter.

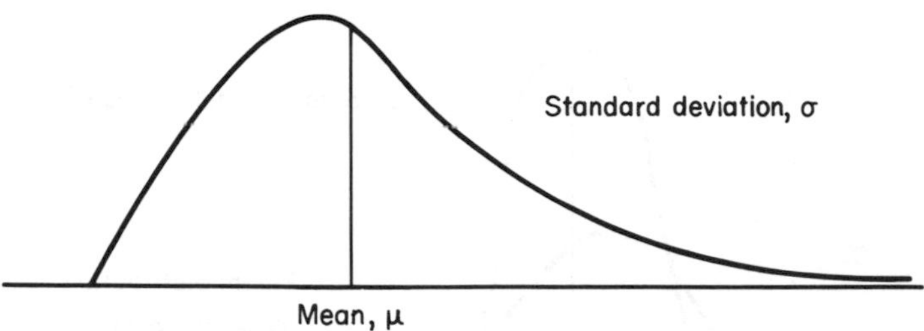

Fig. 8.2 A continuous probability distribution.

8.5 SAMPLING DISTRIBUTION OF THE SAMPLE MEAN

This is not a book about the mathematical theory of statistics, but the theory which we will now discuss is essential for a more complete understanding of the inferential methods to be discussed later, particularly those concerning confidence intervals (in the next chapter).

Suppose that we are sampling from a population of measurements which has a mean, μ, and standard deviation, σ. We will suppose that the measurement is of a continuous variable so that its probability distribution is continuous but not necessarily normal (see Fig. 8.2).

Suppose we select a random sample of size n from this 'parent' population, and calculate $\bar{x}$, the sample mean (using the methods of Section 4.2). If we continue this procedure of taking random samples of size n and calculate the sample mean on each occasion, we will eventually have a large number of such values. Since the random samples are very unlikely to consist of the same values, these sample means will vary and form a population with a distribution which is called the *sampling distribution of the sample mean*.

This distribution will have a mean and standard deviation which we will denote by $\mu_{\bar{x}}$ and $\sigma_{\bar{x}}$ to distinguish them from the μ and σ of the parent population. The sampling distribution will clearly have a shape. The question of interest is, 'What is the connection between the mean, standard deviation and shape of the distribution of the parent population and the mean, standard deviation and shape of the sampling distribution of the sample mean?'

The answer to this question is supplied by some of the most important results in statistical theory which we quote now (without proof!) in three parts:

1. $\mu_{\bar{x}} = \mu$
2. $\sigma_{\bar{x}} = \sigma/\sqrt{n}$
3. If n is large, the distribution of the sample mean is approximately normal, irrespective of the shape of the distribution of the parent

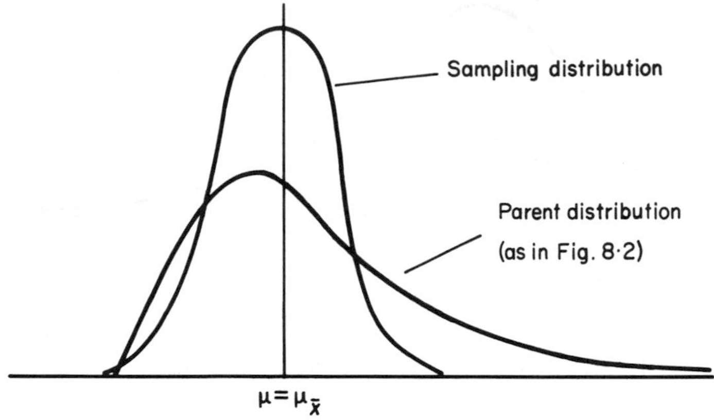

Fig. 8.3 A 'parent' distribution and the sampling distribution of the sample mean.

population. (If, however, the 'parent' distribution is normal then the distribution of the sample mean is also normal for all values of n.)

The beauty of mathematics (if non-mathematicians will allow me to eulogize for just one sentence) is that so much can be stated in so few symbols. We now use a picture and some words to explain the three parts of this theory (see Fig. 8.3).

In words the three parts of the theory state that, if lots of random samples of the same size, n, are taken from a parent distribution, then:

1. The mean of the sample means will equal the mean of the parent.
2. The standard deviation of the sample means will be smaller than the standard deviation of the parent by a factor of $\sqrt{n}$. For example, if $n = 4$, the standard deviation of the sample means will be half the standard deviation of the parent.
3. The shape of the sampling distribution of the sample mean is approximately normal for large sample sizes (and the larger the value of the sample size, the more closely is this distribution normal), even if the parent distribution is not normal.

This third result is called the *central limit theorem* and is an example of one of the two reasons (see Section 7.2) for the importance of the normal distribution in statistics.

In the next chapter we will use the three results quoted above to make estimates of the mean of a population from sample data (these estimates are called confidence intervals).

8.6 SUMMARY

A population is the whole set of measurements about which we want to draw a conclusion, and a sample is a sub-set of a population. Conclusions may be drawn from sample data if an appropriate sampling method and sample size are chosen. Random sampling is important because the theory of how to make inferences about populations from randomly sampled data is well developed. Systematic sampling may sometimes be used instead of random sampling, and stratified sampling is appropriate when the population is, in reality, made up of a number of sub-populations. The required sample size depends on the precision required in the estimate of the population parameter, and on the variability in the population.

The theory of inference depends on the distribution of sample statistics, such as the sample mean. The theory relates the characteristics of the sampling distribution with those of the distribution of the parent population.

WORKSHEET 8: SAMPLES AND POPULATIONS

1. What is (a) A population? (b) A sample? (c) A random sample? (d) A biased sample? (e) A census?

2. Why and how are random samples taken? Think of an example in your main subject area.

3. What is wrong with each of the following methods of sampling? In each case think of a better method.
   (a) In a laboratory experiment with mice, five mice were selected by the investigator by plunging a hand into a cage containing 20 mice and catching them one at a time.
   (b) In a survey to obtain adults' views on unemployment people were stopped by the investigator as they came out of (i) a travel agent, (ii) a food supermarket, (iii) a job centre.
   (c) In a survey to decide whether two species of plant tended to grow close together in a large meadow the investigator went into the meadow and randomly threw a quadrat over his left shoulder a number of times, each time noting the presence or absence of the species (a quadrat is, typically, a metal or plastic frame 1 metre square).
   (d) A survey of the price of bed and breakfast was undertaken in a town with 40 three-star, 50 two-star and 10 one-star hotels. A sample of ten hotels was obtained by numbering the hotels from 00 to 99 and using random number tables.
   (e) To save time in carrying out an opinion poll in a constituency, a party worker selected a random sample from the electoral register for the constituency, and telephoned those selected.
   (f) To test the efficacy of an anti-influenza vaccine in his practice, a doctor placed a notice in his surgery asking patients to volunteer to be given the vaccine.

4. When a die is thrown once, the discrete probability distribution for the number on the uppermost face is:

Number	1	2	3	4	5	6
Probability	$\frac{1}{6}$	$\frac{1}{6}$	$\frac{1}{6}$	$\frac{1}{6}$	$\frac{1}{6}$	$\frac{1}{6}$

The mean and standard deviation of the distribution are $\mu = 3.5$, $\sigma = 1.71$. Now conduct the following experiment: throw two dice and note the mean of the two scores on the uppermost face. Repeat a total of 108 times and form a frequency distribution, noting that the possible values for the mean score are 1, 1.5, 2.0, . . ., 6.0. Calculate the mean and standard deviations of this distribution. Assuming these are reasonably good estimates of $\mu_{\bar{x}}$ and $\sigma_{\bar{x}}$ for samples of size $n = 2$, you should find that $\mu_{\bar{x}} = \mu = 3.5$ and $\sigma_{\bar{x}} = \sigma/\sqrt{2} = 1.21$ (approximately).

Also represent the frequency distribution in a line chart. The shape of this chart should be approximately triangular, which is a little more normal than the flat shape of the parent distribution.

Repeat this experiment for three dice.

Confidence interval estimation

9.1 INTRODUCTION

Suppose we take a random sample of 50 from the population of the heights of all first-year students in higher education, and we find that the sample mean height is $\bar{x} = 171.2$ cm.

What would our estimate be for the population mean height, μ? The obvious answer is 171.2 if we require a single-value estimate (a single-value estimate is referred to as a *point* estimate). However, since our estimate is based only on a sample of the population of heights, we might want to be more guarded and give some idea of the precision of the estimate by adding an 'error term' which might, for example, lead to a statement that the estimate is 171.2 ± 1, meaning that the population mean height, μ, lies somewhere between 170.2 and 172.2 cm. Such an estimate is referred to as an *interval* estimate. The size of the error term will depend on three factors:

1. The sample size, n. The larger the sample size, the better the estimate, the smaller the error term and the greater the precision.
2. The variability in height (as measured by the standard deviation). The larger the standard deviation, the poorer the estimate, the larger the error term and the smaller the precision.
3. The level of confidence we wish to have that the population mean height does in fact lie within the interval specified. This is such an important concept that we will devote the next section to it.

9.2 95% CONFIDENCE INTERVALS

For the student height example, the population mean height, μ, has a fixed numerical value at any given time. We do not know this value, but by taking a random sample of 50 heights we want to specify an interval within which we want to be reasonably confident that this fixed value lies.

Suppose we decide that nothing less than 100% confidence will suffice, implying absolute certainty. Unfortunately, theory indicates that the 100% confidence interval is so wide that it is useless for all practical purposes. Typically investigators choose a 95% confidence level and calculate a 95% confidence interval for the population mean, μ, using formulae we shall introduce in the next sections. For the moment it is important to understand what a confidence level of 95% means. It means that on 95% of occasions when such intervals are calculated the population mean will actually fall inside the calculated interval, but on 5% of occasions it will fall outside the interval. In a particular case, however, we will not know whether the mean has been successfully 'captured' within the calculated interval.

9.3 CALCULATING A 95% CONFIDENCE INTERVAL FOR THE MEAN, μ, OF A POPULATION–LARGE SAMPLE SIZE, n

Having discussed some concepts we now discuss methods of actually calculating a 95% confidence interval for the population mean, μ.

In Section 8.5, we stated that the sample mean has a distribution with a mean, μ, standard deviation $\sigma/\sqrt{n}$, and is approximately normal if n is large (see Fig. 9.1).

Recall that μ and σ were the mean and standard deviation of the 'parent' distribution. It follows from the theory of the normal distribution that 95% of the values of the sample mean lie within 1.96 standard deviations of the mean of the distribution of sample means (refer to Section 7.2 if necessary).

So 95% of the values of the sample mean lie in the range $\mu - 1.96\sigma/\sqrt{n}$ to $\mu + 1.96\sigma/\sqrt{n}$. We can write this as a probability statement:

$$P(\mu - 1.96\sigma/\sqrt{n} \leqslant \bar{x} \leqslant \mu + 1.96\sigma/\sqrt{n}) = 0.95.$$

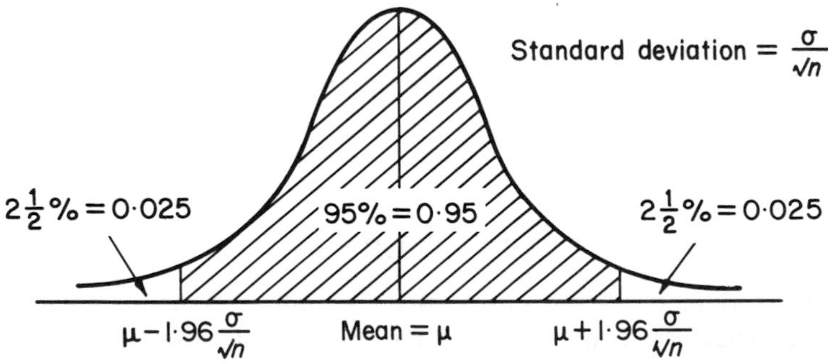

Fig. 9.1 Distribution of the sample mean $\bar{x}$, if n is large. Standard deviation = $\sigma/\sqrt{n}$.

This statement can be rearranged to state:

$$P(\bar{x} - 1.96\sigma/\sqrt{n} \leqslant \mu \leqslant x + 1.96\sigma/\sqrt{n}) = 0.95.$$

If we now use the fact that n is large, we could replace σ† by the sample standard deviation, s, which we calculate from sample data, and obtain the following statement which is now approximately true:

$$P(\bar{x} - 1.96s/\sqrt{n} \leqslant \mu \leqslant \bar{x} + 1.96s/\sqrt{n}) = 0.95.$$

The importance of this result is that we can calculate $\bar{x} - 1.96s/\sqrt{n}$ and $\bar{x} + 1.96s/\sqrt{n}$ from our sample data. These values are called 95% *confidence limits* for μ.

The interval $\bar{x} - 1.96s/\sqrt{n}$ to $\bar{x} + 1.96s/\sqrt{n}$ is called a 95% *confidence interval* for μ. The 'error' term (referred to in Section 9.1) is $1.96s/\sqrt{n}$. Note that we can only use this formula if n is large.

Example

Suppose that from a random sample of 50 heights we find that $\bar{x} = 171.2$, $s = 10$ cm. Then a 95% confidence interval for the population mean height, μ, is:

$$171.2 - 1.96 \times \frac{10}{\sqrt{50}} \quad \text{to} \quad 171.2 + 1.96 \times \frac{10}{\sqrt{50}}$$

$$168.4 \quad \text{to} \quad 174.0.$$

So we are 95% confident that μ lies between 168.4 and 174.0 cm, the error term is 2.8 cm, and the width of the interval is 174.0 − 168.4 = 5.6 cm. See Fig. 9.2.

The formula for the error term is $1.96s/\sqrt{n}$, and this can be used to support the intuitive arguments of Section 9.1 concerning the three factors which affect the 'error' term:

1. Clearly as n increases, $1.96s/\sqrt{n}$ decreases.
2. The larger the variability, the larger s will be, so $1.96s/\sqrt{n}$ will be larger.
3. We use the factor 1.96 in the error term for 95% confidence. For greater

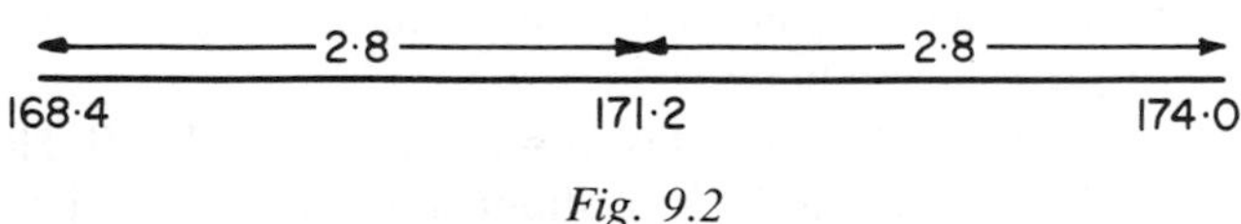

Fig. 9.2

† Some texts discuss the case where σ, the population standard deviation, is known (as an exact value). This author's view is that this is unrealistic since if we know σ, we must have all the measurements in the population and hence we also know μ, the population mean, and there is no need to estimate it.

confidence we would need to move further away from the mean (see Fig. 9.1), so the factor 1.96 would increase and so would the error term. So the greater the confidence level, the greater the error term, and the wider the interval.

9.4 CALCULATING A 95% CONFIDENCE INTERVAL FOR THE MEAN, μ, OF A POPULATION–SMALL SAMPLE SIZE, n

The formula of the previous section referred to cases where the sample size n is large, but 'large' was not defined although $n > 30$ can be used as a rough guide. If we can assume that the variable is approximately normally distributed and $n \geqslant 2$, a 95% confidence interval for μ is given by:

$$\bar{x} - \frac{ts}{\sqrt{n}} \quad \text{to} \quad \bar{x} + \frac{ts}{\sqrt{n}}.$$

Assumption in using this formula The variable, x, must be approximately normally distributed. This assumption is less critical the larger the value of n.

In this formula the value of t is obtained from Table D.5 of Appendix D (percentage points of the t-distribution – see Section 9.5) and depends on two factors:

1. The confidence level, which determines the values of α to be entered in Table D.5. For example, for 95% confidence, $\alpha = (1 - 0.95)/2 = 0.025$.
2. The sample size, n, which determines the number of degrees of freedom, ν, to be entered in Table D.5. The general idea of what is meant by degrees of freedom is discussed in Section 9.7. In estimating the population mean, using the formula given above, $\nu = (n - 1)$.

We also note that the 'error' term in the formula is $ts/\sqrt{n}$.

Example

Suppose the sample data $\bar{x} = 171.2$, $s = 10$ cm had been calculated from a sample of only 10. Since we can safely assume that height is normally distributed (being a measurement on nominally identical individuals) we can use the formula given above.

For 95% confidence, $\alpha = 0.025$. For $n = 10$, $\nu = 10 - 1 = 9$. Hence from Table D.5, $t = 2.262$. Thus a 95% confidence interval for μ is:

$$171.2 - \frac{2.262 \times 10}{\sqrt{10}} \quad \text{to} \quad 171.2 + \frac{2.262 \times 10}{\sqrt{10}}$$

$$164.0 \quad \text{to} \quad 178.4.$$

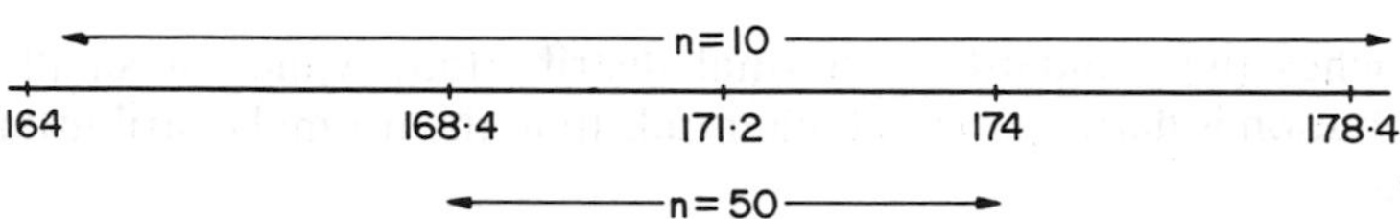

Fig. 9.3 Effect of sample size on a 95% confidence interval for μ.

Based on a sample of 10, we are 95% confident that μ lies between 164.0 and 178.4. The width of the 95% confidence interval has grown from 5.6 cm (for $n = 50$) to 14.4 cm (for $n = 10$). See Fig. 9.3.

Note

If we use the t tables for large values of n, then for 95% confidence, and hence $\alpha = 0.025$, the t value is 1.96 and we have the formula we used in Section 9.3. We may therefore use the formula $\bar{x} \pm ts/\sqrt{n}$ for all $n \geqslant 2$, and stop wondering what 'n must be large' means.

*9.5 THE t-DISTRIBUTION

This continuous probability distribution was first studied by W.S. Gosset, who published his results under the pseudonym of Student, which is why it is often referred to as Student's t-distribution. It arises when we consider taking a large number of random samples of the same size, n, from a normal distribution with known mean, μ. Then the probability distribution of the statistic

$$t = \frac{\bar{x} - \mu}{s/\sqrt{n}}$$

may be plotted. It will be symmetrical and unimodal. For different values of n, different distributions will be obtained; for large n the t-distribution

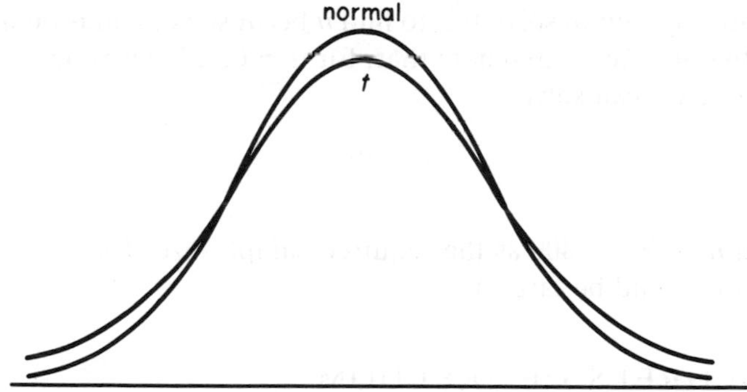

Fig. 9.4 Comparison of the shapes of a normal distribution and a t-distribution with $\nu = 9$ degrees of freedom.

approaches the standardized normal distribution, while for small n the t-distribution is flatter and has higher tails than the normal distribution. See Fig. 9.4.

9.6 THE CHOICE OF SAMPLE SIZE WHEN ESTIMATING THE MEAN OF A POPULATION

In Section 8.4 we discussed the choice of sample size, n, but deferred deciding how to calculate how large it should be until this chapter.

Instead we concentrated on the factors affecting the choice of n for the case of estimating the mean of a population, μ. These factors were:

1. The precision with which the population mean is to be estimated, and we can now state this precision in terms of the 'error' term in the formula for the confidence interval for μ.
2. The variability of the measurements, and we noted a chicken-and-egg situation of needing to know the variability before we had any sample data. To overcome this difficulty we can carry out a pilot experiment and get at least a rough estimate of the standard deviation.

Example

Suppose we specify a precision in terms of an 'error' term of 1 for 95% confidence.

Then $ts/\sqrt{n} = 1$, where t is found from Table D.5 for $\alpha = 0.025$, but n and hence $\nu = (n - 1)$ are unknown.

Suppose we have a rough estimate that $s = 10$ from a small pilot experiment. Then

$$\frac{t \times 10}{\sqrt{n}} = 1.$$

We apparently cannot solve this to find n because t depends on n. But if we assume that n is large, and note that, for $\alpha = 0.025$, t is roughly 2 for large values of n, we can solve

$$\frac{2 \times 10}{\sqrt{n}} = 1$$

to obtain $n = 20^2 = 400$ as the required sample size. (We were correct in assuming n would be large.)

*9.7 DEGREES OF FREEDOM

There are two approaches which you can take to the concept of degrees of freedom when reading this book. The 'cook-book' approach is to know where to find the formula for calculating degrees of freedom for each application covered (and there are quite a few in the remaining chapters). The more mature approach is to try to understand the general principle behind all the formulae for degrees of freedom, which is as follows:

> The number of degrees of freedom equals the number of independent values used in calculating a statistic, minus the number of restrictions placed on the data.

Example

Why do we use $(n - 1)$ degrees of freedom when we look up t in Table D.5 in the calculation of 95% confidence limits for μ? The answer is that in the formula $\bar{x} \pm ts/\sqrt{n}$, we calculate s using

$$s = \sqrt{\frac{\Sigma x^2 - \dfrac{(\Sigma x)^2}{n}}{n - 1}}.$$

An alternative form of this formula (see Section 4.7) is:

$$s = \sqrt{\frac{\Sigma(x - \bar{x})^2}{n - 1}} = \sqrt{\frac{(x_1 - \bar{x})^2 + (x_2 - \bar{x})^2 + \cdots + (x_n - \bar{x})^2}{n - 1}}.$$

Only $(n - 1)$ of the differences $(x_1 - \bar{x})$, $(x_2 - \bar{x})$, etc., are independent since there is the restriction that the sum $\Sigma(x - \bar{x}) = 0$ (we saw examples of this in Worksheet 2, Questions 4 and 5).

9.8 95% CONFIDENCE INTERVAL FOR A BINOMIAL PROBABILITY

All the discussion so far in this chapter has been concerned with confidence intervals for the mean, μ, of a population. If our sample data is from a binomial experiment for which we do not know the value of p, the probability of success in each trial (in other words the proportion of successes in a large number of trials), we can use our sample data to calculate a 95% confidence interval for p. Thus if we have observed x successes in the n trials of a binomial experiment, then a 95% confidence interval for p is

$$\frac{x}{n} \pm 1.96\sqrt{\frac{\dfrac{x}{n}\left(1 - \dfrac{x}{n}\right)}{n}}, \quad \text{provided } x > 5 \quad \text{and} \quad n - x > 5.^{\dagger}$$

† These conditions are the equivalent of $np > 5$ and $n(1 - p) > 5$ for the normal approximation to the binomial (Section 7.6), where the unknown p is replaced by its point estimator, x/n.

Assumption in using this formula The four conditions for the binomial apply (see Section 6.3).

Example

Of a random sample of 200 voters taking part in an opinion poll, 110 said they would vote for Party A, the other 90 said they would vote for other parties. What proportion of the total electorate will vote for Party A?

If we regard 'voting for A' as a 'success' then $x = 110$, $n = 200$. The conditions $x > 5$ and $n - x > 5$ are satisfied, so a 95% confidence interval for p is:

$$\frac{110}{200} \pm 1.96\sqrt{\frac{\frac{110}{200}\left(1 - \frac{110}{200}\right)}{200}} = 0.55 \pm 0.07$$

or 0.48 to 0.62.

We can be 95% confident that the proportion who will vote for Party A is between 0.48 (48%) and 0.62 (62%).

9.9 THE CHOICE OF SAMPLE SIZE WHEN ESTIMATING A BINOMIAL PROBABILITY

In the example of the previous section the width of the confidence interval is large. If we wished to reduce the width by reducing the error term, we would need to increase the sample size.

Example

If we wished to estimate the proportion to within (an error term of) 0.02 for 95% confidence, the new sample size, n, can be found by solving

$$1.96\sqrt{\frac{\frac{110}{200}\left(1 - \frac{110}{200}\right)}{n}} = 0.02.$$

That is

$$n = \frac{1.96^2 \times 0.55 \times 0.45}{0.02^2} = 2377.$$

We need a sample of nearly 2500. Notice how we again used the result of a pilot survey (of 200 voters) as in Section 9.6.

9.10 95% CONFIDENCE INTERVAL FOR THE MEAN OF A POPULATION OF DIFFERENCES, 'PAIRED' SAMPLES DATA

In experimental work we are often concerned with not just one population, but a comparison between two populations.

Example

Two methods of teaching children to read are compared. Some children are taught by a standard method (S) and some by a new method (N). In order to reduce the effect of factors other than teaching method, children are matched in pairs so that the children in each pair are as similar as possible with respect to factors such as age, sex, social background and initial reading ability. One child from each pair is then randomly assigned to teaching method S and the other to method N. Suppose that after one year the children are tested for reading ability, and suppose the data in Table 9.1 are the test scores for 10 pairs of children.

Table 9.1 Reading test scores of 10 matched pairs of children

Pair number	1	2	3	4	5	6	7	8	9	10
S method score	56	59	61	48	39	56	75	45	81	60
N method score	63	57	67	52	61	71	70	46	93	75
d = *N score* − *S score*	7	−2	6	4	22	15	−5	1	12	15

In this example we can think of two populations of measurements, namely the S method scores and the N method scores. However, if our main interest is in the difference between the methods, the one population about which we wish to draw inferences is the population of differences between pairs.

The sample data in Table 9.1 are an example of *paired* samples data. The differences, d, have been calculated in the bottom row of Table 9.1. Then a 95% confidence interval for μ_d, the mean of the population of differences, can be calculated using

$$\bar{d} \pm \frac{ts_d}{\sqrt{n}}$$

where $\bar{d}$ is the sample mean difference, so $\bar{d} = (\Sigma d)/n$, and s_d is the sample standard deviation of differences. So

$$s_d = \sqrt{\frac{\Sigma d^2 - \frac{(\Sigma d)^2}{n}}{n-1}},$$

n is the number of differences (= number of pairs)
t is obtained from Table D.5 for $\alpha = 0.025$, $\nu = (n - 1)$.

Assumption in using this formula The differences must be approximately normally distributed. This assumption is less critical the larger the value of n.

For the data in Table 9.1, $\bar{d} = 7.5$, $s_d = 8.48$, $n = 10$, $t = 2.262$, so a 95% confidence interval for μ_d is

$$7.5 \pm \frac{2.262 \times 8.48}{\sqrt{10}}$$

i.e. 1.4 to 13.6.

We are 95% confident that the mean difference between the methods lies between 1.4 and 13.6, where difference = N score − S score.

As in Section 9.6, it would now be possible to decide what sample size to choose in another experiment designed to provide a more precise estimate of the mean difference between the methods.

9.11 95% CONFIDENCE INTERVAL FOR THE DIFFERENCE IN THE MEANS OF TWO POPULATIONS, 'UNPAIRED' SAMPLES DATA

The example of the previous section was not, in essence, a comparison of two populations since the data were in pairs. In many other instances where two populations of measurements are concerned, the data are *unpaired*.

Example

The heights of a random sample of 50 students in their first year in higher education were summarized as follows:

$$\bar{x}_1 = 171.2 \text{ cm}, \quad s_1 = 10 \text{ cm}, \quad n_1 = 50.$$

Three years later the heights of a second random sample of 40 students in their first year in higher education were summarized as follows:

$$\bar{x}_2 = 173.7 \text{ cm}, \quad s_2 = 9.2 \text{ cm}, \quad n_2 = 40.$$

The suffixes 1 and 2 have been used to denote data from the first and second populations of heights, respectively. The data are unpaired in the sense that no individual height in the first sample is associated with any individual height in the second sample.

We will use these data to estimate the difference between the means of the two populations $(\mu_1 - \mu_2)$. The formula for a 95% confidence interval for $(\mu_1 - \mu_2)$ is

$$(\bar{x}_1 - \bar{x}_2) \pm ts\sqrt{\left(\frac{1}{n_1} + \frac{1}{n_2}\right)}$$

where t is found from Table D.5 for $\alpha = 0.025$, $\nu = (n_1 + n_2 - 2)$ degrees of freedom, and

$$s^2 = \frac{(n_1 - 1)s_1^2 + (n_2 - 1)s_2^2}{n_1 + n_2 - 2}.$$

This formula shows that s^2 is a weighted average of the sample variances s_1^2 and s_2^2, and is called a 'pooled' estimate of the common variance of the two populations – see Assumption 2 below.

Assumptions in using this formula

1. The measurements in each population must be approximately normally distributed, this assumption being less critical the larger the values of n_1 and n_2.
2. The population standard deviations, σ_1 and σ_2, must be equal. Clearly this implies that the population variances, σ_1^2 and σ_2^2, are also equal.

For the numerical example of the height data, we can safely assume normality and, in any case, the sample sizes are large. It is also reasonable to assume the population standard deviations are equal since the values of the sample standard deviations, s_1 and s_2, and in close agreement. Using the formula for s^2, initially,

$$s^2 = \frac{(50 - 1) \times 10^2 + (40 - 1) \times 9.2^2}{50 + 40 - 2} = 93.19$$

$$s = 9.65.$$

95% confidence interval for $(\mu_1 - \mu_2)$ is

$$(171.2 - 173.7) \pm 1.99 \times 9.65\sqrt{\left(\frac{1}{50} + \frac{1}{40}\right)}$$

i.e. -6.6 to $+1.6$.

We are 95% confident that the difference in the mean heights of the two populations is between -6.6 cm and $+1.6$ cm. As in previous examples more precise estimates could have been obtained (if required) by taking larger samples.

9.12 SUMMARY

A confidence interval for a parameter of a population, such as the mean, is a range within which we have a particular level of confidence, such as 95%, that the parameter lies. If we have randomly sampled data we can calculate confidence intervals for various parameters using one of the formulae in Table 9.2.

Table 9.2 Formulae for 95% confidence intervals

Parameter	*Case*	*Assumption*	*Formula*
μ Population mean	n large		$\bar{x} \pm 1.96\frac{s}{\sqrt{n}}$
μ Population mean	$n \geqslant 2$	Variable approximately normal	$\bar{x} \pm \frac{ts}{\sqrt{n}}$
p Binomial probability	$x > 5$ $n - x > 5$	The four conditions for a binomial distribution	$\frac{x}{n} \pm 1.96\sqrt{\frac{\frac{x}{n}\left(1 - \frac{x}{n}\right)}{n}}$
μ_d The mean of a population of differences	$n \geqslant 2$	Differences approximately normal	$\bar{d} \pm \frac{ts_d}{\sqrt{n}}$
$\mu_1 - \mu_2$ The difference in means of two populations	$n_1, n_2 \geqslant 2$	(i) Variable approximately normal	$(\bar{x}_1 - \bar{x}_2) \pm ts\sqrt{\left(\frac{1}{n_1} + \frac{1}{n_2}\right)}$, where
		(ii) $\sigma_1 = \sigma_2$	$s^2 = \frac{(n_1 - 1)s_1^2 + (n_2 - 1)s_2^2}{n_1 + n_2 - 2}$

We can decide sample sizes if we can specify the precision with which we wish to estimate the parameter and if we have some measure of variability from the results of a pilot experiment or survey.

WORKSHEET 9: CONFIDENCE INTERVAL ETIMATION

1. Why are confidence intervals calculated?
2. What information do we need to calculate a 95% confidence interval for the mean of a population? What assumption is required if the sample size is small?

3. The larger the sample size, the wider the 95% confidence interval. True or false?

4. The more variation in the measurements, the wider the 95% confidence interval. True or false?

5. The higher the level of confidence, the wider the confidence interval. True or false?

6. What does the following statement mean: 'I am 95% confident that the mean of the population lies between 10 and 12.'

7. Of a random sample of 100 customers who had not settled accounts with an Electricity Board within one month of receiving them, the mean amount owed was £30 and the standard deviation was £10. What is your estimate of the mean of all unsettled accounts? Suppose the Electricity Board wanted an estimate of the mean of all unsettled accounts to be within £1 of the true figure for 95% confidence. How many customers who had not settled accounts would need to be sampled?

8. Out of a random sample of 100 people, 80 said they were non-smokers. Estimate the percentage of non-smokers in the population with 95% confidence. How many people would need to be asked if the estimated percentage of non-smokers in the population is required to be within 1% for 95% confidence?

9. The systolic blood pressure of 90 normal British males has a mean of 128.9 mm of mercury and a standard deviation of 17 mm of mercury. Assuming these are a random sample of blood pressures, calculate a 95% confidence interval for the population mean blood pressure.
   (a) How wide is the interval?
   (b) How wide would the interval be if the confidence level was raised to 99%?
   (c) How wide would the 95% confidence interval be if the sample size was increased to 360?

   Are your answers to (a), (b) and (c) consistent with your answers to Questions 3 and 5 above?

10. In order to estimate the percentage of pebbles made of flint in a given locality to within 1% for 95% confidence, a pilot survey was carried out. Of a random sample of 30 pebbles, 12 were made of flint. How many pebbles need to be sampled in the main survey?

11. The number of drinks sold from a vending machine in a motorway service station was recorded on 60 consecutive days. The results were summarized in a grouped frequency distribution table:

Number of drinks	*Number of days*
0– 99	4
100–199	8
200–299	15
300–399	20
400–499	9
500–599	3
600–699	1

Ignoring any differences between different days of the week and any time-trend or seasonal effects, estimate the mean number of drinks sold per day in the long term.

12. Ten women recorded their weights in kilograms before and after dieting. Assuming the women were randomly selected, estimate the population mean reduction in weight. What additional assumption is required and is it reasonable to make it here?

Table 9.3 Weights (kg) before and after dieting for 10 women

Before	89.1	68.3	77.2	91.6	85.6	83.2	73.4	84.3	96.4	87.6
After	84.3	66.2	76.8	79.3	85.5	80.2	76.2	80.3	90.5	80.3

13. The percentage of a certain element in an alloy was determined for 16 specimens using one of two methods A and B. Eight of the specimens were randomly allocated to each method. The percentage were:

Method A	13.3	13.4	13.3	13.5	13.6	13.4	13.3	13.4
Method B	13.9	14.0	13.9	13.9	13.9	13.9	13.8	13.7

Calculate a 95% confidence interval for the difference in the mean percentage of the element in the alloy for the two methods, stating any assumptions made.

14. The annual rainfall in cm in two English towns over 11 years was as follows:

Year	*Town A*	*Town B*
1970	100	120
1971	89	115
1972	84	96
1973	120	115
1974	130	140
1975	105	120
1976	60	75
1977	70	90
1978	90	90
1979	108	105
1980	130	135

Estimate the mean difference in the annual rainfall for the two towns.

15. The actual weights of honey in 12 jars marked 452 g were recorded. Six of the jars were randomly selected from a large batch of Brand A honey, and six were randomly selected from a large batch of Brand B honey. The weights were:

Brand A	442	445	440	448	443	450
Brand B	452	450	456	456	460	449

Estimate the mean difference in the weights of honey in jars marked 452 g for the two brands. Also estimate separately
(a) the mean weight of Brand A honey, and
(b) the mean weight of Brand B honey.

Decide whether it is reasonable to suppose that the mean weight of honey from the Brand A batch is 452 g, and similarly for Brand B honey.

Hypothesis testing

What tribunal can possibly decide truth in the clash of contradictory assertions and conjectures?

10.1 INTRODUCTION

Statistical inference is concerned with how we draw conclusions from sample data about the larger population from which the sample is selected. In the previous chapter we discussed one branch of inference, namely estimation, particularly confidence interval estimation. Another important branch of inference is hypothesis testing which is the subject of much of the remainder of this book. See Fig. 10.1.

In this chapter we consider all the cases we looked at in the previous chapter in terms of testing hypotheses about the various parameters, and conclude by discussing the connection between the two branches of inference.

The method of carrying out any hypothesis test can be set out in terms of seven steps:

1. Decide on a null hypothesis, H_0.
2. Decide on an alternative hypothesis, H_1.
3. Decide on a significance level.
4. Calculate the appropriate test statistic.

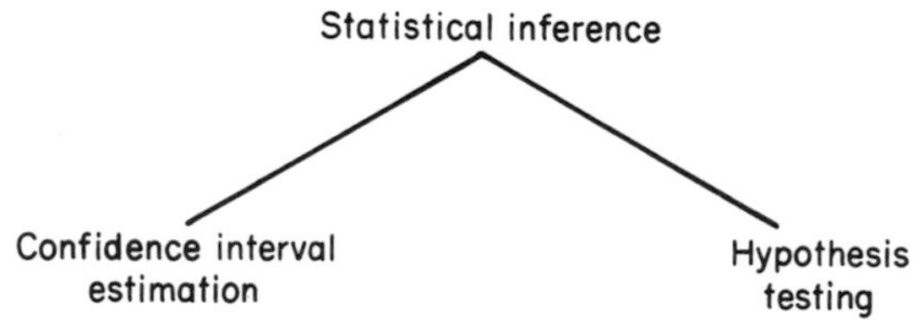

Fig. 10.1

5. Find from tables the appropriate tabulated test statistic.
6. Compare the calculated and tabulated test statistics, and decide whether to reject the null hypothesis, H_0.
7. State a conclusion, and state the assumptions of the test.

In the following sections each of these steps and the concepts behind them will be explained, with the aid of an example.

10.2 WHAT IS A HYPOTHESIS?

In terms of the examples of the previous chapter an hypothesis is a statement about the value of a population parameter, such as a population mean, μ. We use the sample data to decide whether the stated value of the parameter is reasonable. If we decide it is not reasonable we reject the hypothesis in favour of another hypothesis. It is important to note at this stage, then, that in hypothesis testing we have two hypotheses to consider. Using sample data we decide which hypothesis is the more reasonable. We call the two hypotheses the *null hypothesis* and the *alternative hypothesis*.

10.3 WHICH IS THE NULL HYPOTHESIS AND WHICH IS THE ALTERNATIVE HYPOTHESIS?

The null hypothesis generally expresses the idea of 'no difference' – think of 'null' as meaning 'no'. In terms of the examples of the previous chapter a null hypothesis is a statement that the value of the population parameter is 'no different from', that is 'equal to', a specified value. The symbol we use to denote a null hypothesis is H_0.

Example of a null hypothesis

$$H_0 : \mu = 175.$$

This null hypothesis states that the population mean equals 175.

The alternative hypothesis, which we denote by H_1, expresses the idea of 'some difference'. Alternative hypotheses may be *one-sided* or *two-sided*. The first two examples below are one-sided since each specifies only one side of the value 175; the third example is two-sided since both sides of 175 are specified.

Examples of alternative hypotheses

$H_1 : \mu > 175$ Population mean greater than 175,

$H_1 : \mu < 175$ Population mean less than 175,

$H_1 : \mu \neq 175$ Population mean not equal to 175.

In a particular case we specify both the null and alternative hypotheses according to the purpose of the investigation, and before the sample data are collected.

10.4 WHAT IS A SIGNIFICANCE LEVEL?

Hypothesis testing is also sometimes referred to as *significance testing*. The concept of *significance level* is similar to the concept of confidence level. The usual value we choose for our significance level is 5%, just as we usually choose a confidence level of 95%. Just as the confidence level expresses the idea that we would be prepared to bet heavily that the interval we state actually does contain the value of the population parameter of interest (see Worksheet 9, Question 6 and its solution), so a significance level of 5% expresses a similar idea in connection with hypothesis testing:

> A significance level of 5% is the risk we take in rejecting the null hypothesis, H_0, in favour of the alternative hypothesis, H_1, when in reality H_0 is the correct hypothesis.

Example

If the first three steps (see Section 10.1) of our hypothesis test are:

1. $H_0: \mu = 175$,
2. $H_1: \mu \neq 175$,
3. 5% significance level,

then we are stating that we are prepared to run a 5% risk that we will reject H_0 and conclude that the mean is not equal to 175, when the mean actually is equal to 175.

We cannot avoid the small risk of drawing such a wrong conclusion in hypothesis testing, because we are trying to draw conclusions about a population using only part of the information in the population, namely the sample data. The corresponding risk in confidence interval estimation is the small risk that the interval we state will not contain the population parameter of interest.

10.5 WHAT IS A TEST STATISTIC, AND HOW DO WE CALCULATE IT?

A test statistic is a value we can calculate from our sample data and the value we specify in the null hypothesis, using an appropriate formula.

Example

If the first three steps of our hypothesis test are as in the example of Section 10.4, and our sample data are summarized as

$$\bar{x} = 171.2, \quad s = 10, \quad n = 10,$$

then the fourth step of our hypothesis test is as follows:

4. Calculated test statistic is

$$Calc\ t = \frac{\bar{x} - \mu}{s/\sqrt{n}},$$

where μ refers to the value specified in the null hypothesis. (This formula for t was introduced in Section 9.5.)

$$= \frac{171.2 - 175}{10/\sqrt{10}}$$

$$= -1.20.$$

10.6 HOW DO WE FIND THE TABULATED TEST STATISTIC?

We must know which tables to use for a particular application, and how to use them.

Example

1. $H_0 : \mu = 175.$
2. $H_1 : \mu \neq 175.$
3. 5% significance level.
4. $Calc\ t = -1.20$, if $\bar{x} = 171.2$, $s = 10$, $n = 10$.
5. The appropriate table is Table D.5 and we enter the tables for

$$\alpha = \text{significance level}/2,$$

dividing by 2 since H_1 is two-sided,

$$= 0.025,$$

and $\nu = (n - 1) = 9$ degrees of freedom.
So tabulated test statistic is $Tab\ t = 2.262$, from Table D.5.

10.7 HOW DO WE COMPARE THE CALCULATED AND TABULATED TEST STATISTIC?

Example

For the example in Section 10.6 we reject H_0 if $|Calc\ t| > Tab\ t$, where the vertical lines mean that we ignore the sign of $Calc\ t$ and consider only its magnitude (e.g. $|-5| = 5$, $|5| = 5$).

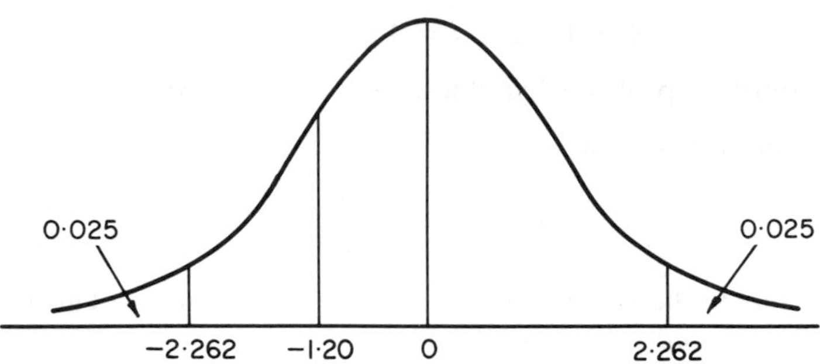

Fig. 10.2 t-distribution for $\nu = 9$ degrees of freedom.

Since in this example $|Calc\ t| = 1.20$, and $Tab\ t = 2.262$, we do not reject H_0. Figure 10.2 shows that only calculated values of t in the tails of the distribution, beyond the *critical values* of -2.262 and 2.262, lead to rejection of H_0.

10.8 WHAT IS OUR CONCLUSION, AND WHAT ASSUMPTIONS HAVE WE MADE?

Our conclusion should be a sentence in words, as far as possible devoid of statistical terminology. For the example above, since we decided not to reject H_0 in favour of an alternative stating that the mean differed from 175, we conclude that 'the mean is not significantly different from 175 (5% level of significance)'.

The assumption of this test is that the variable is approximately normally distributed. This assumption is less critical the larger the sample size.

10.9 HYPOTHESIS TEST FOR THE MEAN, μ, OF A POPULATION

This section is a summary of the seven-step method in terms of the numerical example referred to in Section 10.3 to Section 10.8.

Example

1. $H_0 : \mu = 175$.
2. $H_1 : \mu \neq 175$.
3. 5% significance level.
4.

$$Calc\ t = \frac{\bar{x} - \mu}{s/\sqrt{n}}$$

$$= \frac{171.2 - 175}{10},$$

for sample data $\bar{x} = 171.2$, $s = 10$, $n = 10$,

$$= -1.20.$$

5. *Tab* $t = 2.262$ for

$$\alpha = \frac{\text{sig. level}}{2} = \frac{0.05}{2} = 0.025$$

$$\nu = (n - 1) = 10 - 1 = 9 \text{ degrees of freedom.}$$

6. Since $|Calc\ t| < Tab\ t$, do not reject H_0.
7. Mean is not significantly different from 175 (5% level).
   Assumption Variable is approximately normally distributed.

Warning Note

Notice that although we did not reject H_0, neither did we state that '$\mu = 175$'. We cannot be so definite given that we have only sample data (μ refers to the population). The conclusion in Step 7 simply implies that H_0 is a more reasonable hypothesis than H_1 in this example.

Put another way, there are many other values which could be specified in H_0 which we would not wish to reject, 174 for example. Clearly we cannot state '$\mu = 175$' and '$\mu = 174$', and so on.

10.10 TWO EXAMPLES OF HYPOTHESIS TESTS WITH ONE-SIDED ALTERNATIVE HYPOTHESES

If we had chosen a one sided H_1 in the previous example, the steps would have varied a little. Since we could have chosen $\mu > 175$ or $\mu < 175$ as our alternative, both these examples are now given side by side below:

1.	$H_0 : \mu = 175$	$H_0 : \mu = 175$
2.	$H_1 : \mu > 175$	$H_1 : \mu < 175$
3.	5% significance level	5% significance level
4.	*Calc* $t = -1.20$	*Calc* $t = -1.20$
5.	*Tab* $t = 1.833$ for $\alpha = \frac{0.05}{1} = 0.05$ and $\nu = (n-1) = 9$	*Tab* $t = 1.833$ for $\alpha = \frac{0.05}{1} = 0.05$ and $\nu = (n-1) = 9$
6.	Since *Calc* $t <$ *Tab* t, do not reject H_0	Since *Calc* $t > -$ *Tab* t, do not reject H_0
7.	Mean is not significantly greater than 175 (5% level)	Mean is not significantly less than 175 (5% level)

Assumption	Variable is approximately normally distributed Fig. 10.3 shows that only calculated values of t in the right-hand tail, greater than the critical value of 1.833, lead to rejection of H_0.	Variable is approximately normally distributed Fig. 10.3 shows that only calculated values of t in the left-hand tail, less than the critical value of -1.833, lead to rejection of H_0.

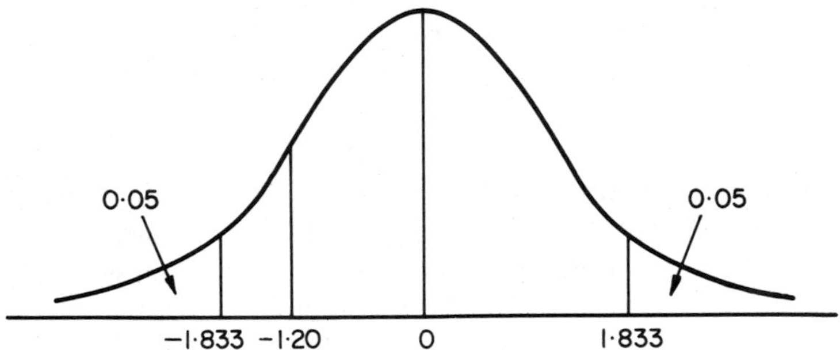

Fig. 10.3 t-distribution for $\nu = 9$ degrees of freedom.

10.11 HYPOTHESIS TEST FOR A BINOMIAL PROBABILITY

Suppose we wish to test a hypothesis for p, the probability of success in a single trial, using sample data from a binomial experiment. If the experiment consisted of n trials, x of which were 'successes', the test statistic is calculated using:

$$Calc\ z = \frac{\frac{x}{n} - p}{\sqrt{\frac{p(1-p)}{n}}},$$

where p is the value specified in the null hypothesis.

We can use this formula if $np > 5$, $n(1 - p) > 5$. The tabulated test statistic is *Tab z*, obtained from Table D.3(b).

Example

Test the hypothesis that the percentage of voters who will vote for Party A in an election is 50% against the alternative that it is greater than 50%, using the random sample data from an opinion poll that 110 out of 200 voters said they would vote for Party A.

1. $H_0: p = 0.5$, which implies 50% vote for Party A.
2. $H_1: p > 0.5$, which implies Party A has an overall majority.
3. 5% significance level.
4.

$$Calc\ z = \frac{\dfrac{x}{n} - p}{\sqrt{\dfrac{p(1-p)}{n}}}$$

can be used since $np = 200 \times 0.5$ and $n(1-p) = 200(1-0.5)$ are both greater than 5

$$= \frac{\dfrac{110}{200} - 0.5}{\sqrt{\dfrac{0.5(1-0.5)}{200}}}$$

$$= 1.414.$$

5. *Tab* $z = 1.645$, since in Table D.3(b) this value of z corresponds to a tail of 0.05/1, the significance level divided by 1, because H_1 is one-sided – see Fig. 10.4.
6. Since *Calc* $z <$ *Tab* z, do not reject H_0.
7. The percentage of voters for Party A is not significantly greater than 50% (5% level); so it is not reasonable to assume that Party A will gain an overall majority in the election.
   Assumption The four binomial conditions apply (see Section 6.3)

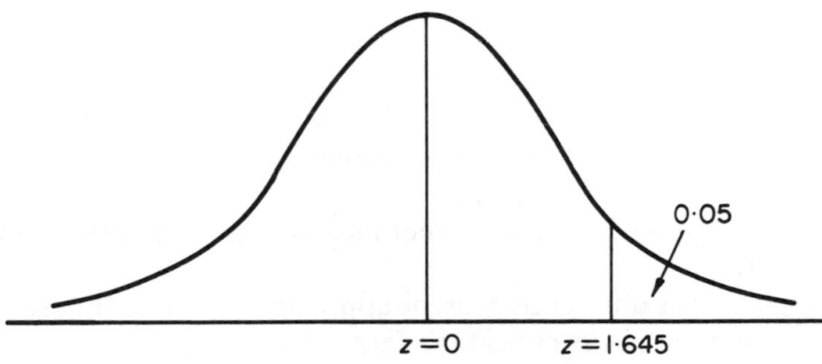

Fig. 10.4 Standardized normal distribution.

10.12 HYPOTHESIS TEST FOR THE MEAN OF A POPULATION OF DIFFERENCES, 'PAIRED' SAMPLES DATA

Example

For the example given in Section 9.10 comparing two methods of teaching children to read, suppose we want to decide whether the new method, N, is better than the standard method, S, in terms of mean difference in the test scores of the two methods.

We assume we have the same data as in Table 9.1, and hence the same summary of that data: $\bar{d} = 7.5$, $s_d = 8.48$, $n = 10$.

1. $H_0 : \mu_d = 0$ implies the mean population difference is zero. That is, the methods give the same mean test score.
2. $H_1 : \mu_d > 0$. Since differences were calculated using N score − S score, this implies that the N method gives a higher mean test score.
3. 5% significance level.
4. $$Calc\ t = \frac{\bar{d}}{s_d/\sqrt{n}},$$

   this is the appropriate test statistic for this 'paired samples *t*-test'

   $$= \frac{7.5}{8.48/\sqrt{10}}$$

   $$= 2.80.$$

5. *Tab t* = 1.833, for

   $$\alpha = \frac{\text{Sig. level}}{1},$$

   since H_1 is one-sided

   $$= 0.05$$

   and $\nu = (n - 1) = 9$ degrees of freedom.
6. Since *Calc t* > *Tab t*, reject H_0.
7. N method gives a significantly higher mean test score than the S method (5% level).

   Assumption The differences must be approximately normally distributed (this assumption is less critical the larger the value of n).

10.13 HYPOTHESIS TEST FOR THE DIFFERENCE IN THE MEANS OF TWO POPULATIONS, 'UNPAIRED' SAMPLES DATA

Example

For the example given in Section 9.11 comparing the heights of two groups of students, suppose we want to decide whether there is a significant difference between the mean heights of the two populations. We assume we have the same data as before, which were summarized as

$$\bar{x}_1 = 171.2 \text{ cm}, \quad s_1 = 10 \text{ cm}, \quad n_1 = 50$$

$$\bar{x}_2 = 173.7 \text{ cm}, \quad s_2 = 9.2 \text{ cm}, \quad n_2 = 40.$$

1. $H_0: \mu_1 = \mu_2$. There is no difference in the population mean heights.
2. $H_1: \mu_1 \neq \mu_2$. There is a difference, in one direction or the other.
3. 5% significance level.
4. $$Calc\ t = \frac{\bar{x}_1 - \bar{x}_2}{s\sqrt{\left(\frac{1}{n_1} + \frac{1}{n_2}\right)}}.$$

   This is the appropriate test statistic for this 'unpaired samples t-test', where

   $$s^2 = \frac{(n_1 - 1)s_1^2 + (n_2 - 1)s_2^2}{n_1 + n_2 - 2}$$

   (s^2 is called a pooled estimate of variance; see also Section 9.11). Using the data above, $s = 9.65$, and

   $$Calc\ t = \frac{171.2 - 173.7}{9.65\sqrt{\left(\frac{1}{50} + \frac{1}{40}\right)}} = -1.22.$$

5. *Tab* $t = 1.99$ for $\alpha = 0.05/2 = 0.025$ and $\nu = n_1 + n_2 - 2 = 88$ degrees of freedom.
6. Since $|Calc\ t| < Tab\ t$, do not reject H_0.
7. Mean heights are not significantly different (5% level).

   Assumptions (1) The measurements in each population must be approximately normally distributed, this assumption being less critical the larger the values of n_1 and n_2. (2) The population standard deviations, σ_1 and σ_2, must be equal. (For these data, both assumptions are reasonable, as discussed in Section 9.11, since the same assumptions applied to the formula there for confidence intervals.)

*10.14 THE EFFECT OF CHOOSING SIGNIFICANCE LEVELS OTHER THAN 5%

Why do we not choose a significance level lower than 5% since we would then run a smaller risk of rejecting H_0 when H_0 is correct (refer to Section 10.4 if necessary)? Just as there are advantages and disadvantages of choosing a confidence level above 95% – a consequence of a higher confidence level is a wider confidence interval – a similar argument applies to a significance level below 5%.

If we reduce the significance level below 5% we reduce the risk of wrongly rejecting H_0, but we increase the risk of drawing a different wrong conclusion, namely the risk of wrongly rejecting H_1. Nor can we set both risks at 5% for the examples described in this chapter (for reasons which are beyond the scope of this book), and even if we could it might not be a wise thing to do! Consider the risks in a legal example and judge whether they should be equal:

(a) The risk of convicting an innocent man in a murder trial.
(b) The risk of releasing a guilty man in a murder trial.

There is nothing sacred about the '5%' for a significance level, nor the '95%' for a confidence level, but we should be aware of the consequences of departing from these conventional levels.

10.15 WHAT IF THE ASSUMPTIONS OF AN HYPOTHESIS TEST ARE NOT VALID?

If at least one of the assumptions of a hypothesis test is not valid, i.e. there is insufficient evidence to make us believe they are all reasonable assumptions, then the test is also invalid and the conclusions may well be invalid.

In such cases, alternative tests called distribution-free tests or more commonly non-parametric tests, should be used. These do not require as many assumptions as the tests described in this chapter, but they have the disadvantages that they are less powerful, meaning that we are less likely to accept the alternative hypothesis when the alternative hypothesis is correct. Some non-parametric tests are described in Chapter 11.

10.16 THE CONNECTION BETWEEN CONFIDENCE INTERVAL ESTIMATION AND HYPOTHESIS TESTING

Confidence interval estimation and hypothesis testing provide similar types of information. However, a confidence interval (if a formula exists to calculate it) provides more information than its corresponding hypothesis test.

Example

Consider the student height data used in the example in Section 9.4. Given sample data $\bar{x} = 171.2$, $s = 10$, $n = 10$, we calculated a 95% confidence interval for the population mean height, μ, as 164.0 to 178.4.

From this we can immediately state that any null hypothesis specifying a value of μ within this interval would not be rejected in favour of the two-sided alternative, assuming a 5% level of significance.

So, for example, $H_0 : \mu = 170$ would not be rejected in favour of $H_1 : \mu \neq 170$ but $H_0 : \mu = 160$ would be rejected in favour of $H_1 : \mu \neq 160$, $H_0 : \mu = 180$ would be rejected in favour of $H_1 : \mu \neq 180$.

We can state that 'a confidence interval for a parameter contains a range of values for the parameter we would not wish to reject'. The confidence interval is a way of representing all the null hypothesis values we would not wish to reject, on the evidence of the sample data.

10.17 SUMMARY

A statistical hypothesis is often a statement about the parameter of a population. In a seven-step method we use sample data to decide whether

Table 10.1 Hypothesis test

Parameter	*Case*	*Assumption*	*Decision rule for two-sided alternative hypothesis: Reject H_0 if*
μ Population mean	$n \geqslant 2$	Variable approximately normal	$\lvert Calc\ t \rvert = \dfrac{\lvert \bar{x} - \mu \rvert}{s/\sqrt{n}} > Tab\ t$
p Binomial probability	$np > 5$ $n(1-p) > 5$	The four conditions for a binomial distribution	$\lvert Calc\ z \rvert = \dfrac{\left\lvert \dfrac{x}{n} - p \right\rvert}{\sqrt{\dfrac{p(1-p)}{n}}} > Tab\ z$
μ_d The mean of a population of differences	$n \geqslant 2$	Differences approximately normal	$\lvert Calc\ t \rvert = \dfrac{\lvert \bar{d} \rvert}{s_d/\sqrt{n}} > Tab\ t$
$\mu_1 - \mu_2$ The difference in means of two populations	$n_1, n_2 \geqslant 2$	(i) Variable is approximately normal (ii) $\sigma_1 = \sigma_2$	$\lvert Calc\ t \rvert = \dfrac{\lvert \bar{x}_1 - \bar{x}_2 \rvert}{s\sqrt{\left(\dfrac{1}{n_1} + \dfrac{1}{n_2}\right)}} > Tab\ t$

to reject the null hypothesis in favour of an alternative hypothesis. Table 10.1 summarizes the various tests covered in this chapter.

If the assumptions of a test are not valid, alternative non-parametric tests (to be discussed in Chapter 11) may be available.

The connection between confidence interval estimation and hypothesis testing was discussed: the former provides more information than the latter.

WORKSHEET 10: HYPOTHESIS TESTING, SOME t-TESTS AND A z-TEST

1. What is
   (a) A (statistical) hypothesis?
   (b) A null hypothesis?
   (c) An alternative hypothesis?
   Give an example of each in your main subject area.
2. What is a significance level?
3. Why do we need to run a risk of wrongly rejecting the hypothesis?
4. Why do we choose 5% as the risk of wrongly rejecting the hypothesis?
5. How can we tell whether an alternative hypothesis is one-sided or two-sided?
6. How do we know whether to specify a one-sided or a two-sided alternative hypothesis in a particular investigation? Think of an example where each would be appropriate.

In Questions 7–18 inclusive, use a 5% significance level unless otherwise stated. In each question the assumptions required for the hypothesis test used should be stated, and you should also decide whether the assumptions are likely to be valid.

7. Eleven cartons of sugar, each nominally containing 1 kg, were randomly selected from a large batch of cartons. The weights of sugar were:

   1.02 1.05 1.08 1.03 1.00 1.06 1.08 1.01 1.04 1.07 1.00.

   Do these data support the hypothesis that the mean weight for the batch is 1 kg?
8. A cigarette manufacturer claims that the mean nicotine content of a brand of cigarettes is 0.30 milligrams per cigarette. An independent consumers group selected a random sample of 1000 cigarettes and found that the sample mean was 0.31 milligrams per cigarette, with a standard deviation of 0.03 milligrams. Is the manufacturers claim

justified or is the mean nicotine content significantly higher than he states?

9. The weekly take-home pay of a random sample of 30 farm workers is summarized in the following table:

Weekly take-home pay (£)	*Number of farm workers*
From 60 up to but not including 70	2
From 70 up to but not including 80	5
From 80 up to but not including 90	9
From 90 up to but not including 100	11
From 100 up to but not including 110	3

Do these data support the claim that the mean weekly take-home pay of farm workers is below £90?

10. The market share for the E10 size packet of a particular brand of washing powder has averaged 30% for a long period. Following a special advertising campaign it was discovered that, of a random sample of 100 people who had recently bought the E10 size packet, 35 had bought the brand in question. Has its market share increased?

11. Of a random sample of 300 gourmets, 180 prefer thin soup to thick soup. Is it reasonable to expect that 50% of all gourmets prefer thin soup?

12. It is known from the records of an insurance company that 14% of all males holding a certain type of life insurance policy die during their 60th, 61st or 62nd year. The records also show that of a randomly selected group of 1000 male civil servants holding this type of policy 112 died during their 60th, 61st or 62nd year. Is it reasonable to suppose that male civil servants have a death rate lower than 14% during these years?

13. An experiment was conducted to compare the performance of two varieties of wheat, A and B. Seven farms were randomly chosen for the experiment, and the yields (in tonnes per hectare) for each variety on each farm were as follows:

Farm number	1	2	3	4	5	6	7
Yield of Variety A	4.6	4.8	3.2	4.7	4.3	3.7	4.1
Yield of Variety B	4.1	4.0	3.5	4.1	4.5	3.2	3.8

(a) Why do you think both varieties were tested on each farm, rather than testing Variety A on seven farms and Variety B on seven other farms?
(b) Carry out a hypothesis test to test whether the mean yields are the same for the two varieties.

14. A sample length of material was cut from each of five randomly selected rolls of cloth and each length divided into two halves. One half was dyed with a newly developed dye, and the other half with a dye that had been in use for some time. The ten pieces were then washed and the amount of dye washed out was recorded for each piece as follows:

Roll	1	2	3	4	5
Old Dye	13.2	13.7	15.4	13.5	16.8
New Dye	12.5	14.3	16.8	14.9	17.4

Investigate the allegation that the amount of dye washed out for the old dye is significantly less than for the new dye.

15. A sleeping drug and a neutral control were tested in turn on a random sample of 10 patients in a hospital. The data below represent the differences between the number of hours of sleep under the drug and the neutral control for each patient:

$$2.0 \quad 0.2 \quad -0.4 \quad 0.3 \quad 0.7 \quad 1.2 \quad 0.6 \quad 1.8 \quad -0.2 \quad 1.0.$$

Is it reasonable to assume that the drug would give more hours of sleep on average than the control for all the patients in the hospital?

16. To compare the strengths of two cements, six mortar cubes were made with each cement and the following strengths were recorded:

Cement A	4600	4710	4820	4670	4760	4480
Cement B	4400	4450	4700	4400	4170	4100

Is there a significant difference between the mean strengths of the two cements?

17. The price of a standard 'basket' of household goods was recorded for 25 corner shops selected at random from the suburbs of all the cities in England. In addition 25 food supermarkets were also randomly selected from the main shopping centres of the cities. The data were summarized as follows:

	Corner shop	*Supermarket*
Mean price (£)	9.45	8.27
Standard deviation of price (£)	0.60	0.50
Number in sample	25	25

Is it reasonable to suppose that corner shops are charging more on average than supermarkets for the standard basket?

18. Two geological sites were compared for the amount of vanadium (ppm) found in clay. It was thought that the mean amount would be less for area A than for area B. Do the following data, which consist of 65 samples taken randomly from each area, support this view?

Amount of vanadium (ppm)	*No. of samples from area A*	*No. of samples from area B*
40–59	4	3
60–79	17	7
80–99	24	14
100–119	12	25
120–139	6	12
140–159	2	4

19. Use the 95% confidence interval method on the data in Questions 7, 11, 13 and 16 to check the conclusions you reached using the hypothesis testing method.

Non-parametric hypothesis tests

At least the difference is very inconsiderable.

11.1 INTRODUCTION

There are some hypothesis tests which do not require as many assumptions as the tests described in Chapter 10. However, these 'non-parametric' tests are less powerful than the corresponding 'parametric' tests, that is we are less likely to reject the null hypothesis and hence accept the alternative hypothesis, when the alternative hypothesis is correct. We can therefore conclude that parametric tests are generally preferred if their assumptions are valid.

Table 11.1

Non-parametric test	*Application*	*Reference for roughly equivalent t-test*
Sign test	Median of a population	Section 10.9
Sign test	Median of a population of differences – 'paired' samples data	Section 10.12
Wilcoxon signed rank test	Median of a population of differences – 'paired' samples data	Section 10.12
Mann–Whitney U test	Difference between the medians of two populations – 'unpaired' samples data	Section 10.13

Three non-parametric tests are described in this chapter for which there are roughly equivalent t-tests, as indicated in Table 11.1.

11.2 SIGN TEST FOR THE MEDIAN OF A POPULATION

Example

Suppose we collect the weekly incomes of a random sample of ten self-employed window cleaners, and we wish to test the hypothesis that the median income is £90 per week for self-employed window cleaners. (We might have started with the idea of testing the hypothesis that the *mean* income is £90 per week, but a cross-diagram of the data below would indicate that the assumption of normality is not reasonable, and hence a t-test would not be a valid test.)

Suppose the incomes are

(£) 75 67 120 62 67 150 65 80 55 90.

The sign test for the median of a population requires that we simply write down the signs of the differences between the incomes and the hypothesized value of 90. Using the convention that incomes below £90 have a − sign, and incomes above £90 have a + sign, and incomes equal to £90 (generally called 'tics') are disregarded, the signs are: − − + − − + − − −.

There are two + and seven − signs. If the null hypothesis that the median is £90 is true, the probability of an income above (or below) £90 is 1/2. We will take the alternative hypothesis to be that the median is not £90. The seven-step method for the sign test for this example is:

1. $H_0: p(+) = p(-) = 1/2$, where $p(+)$ and $p(-)$ mean the probability of a + sign and a − sign, respectively.
2. $H_1: p(+) \neq p(-)$, a two-sided alternative hypothesis.
3. 5% significance level.
4. For the sign test, the calculated test statistic is a binomial probability. In general, it is the probability of getting the result obtained or a result which is more extreme, assuming, for the purposes of calculating the probability, that H_0 is true. For the example we need to calculate:

   P(7 or more minus signs in 9 trials for $p(-) = \frac{1}{2}$)

   $=P(7) + P(8) + P(9)$

   $$= \binom{9}{7}\left(\frac{1}{2}\right)^7\left(1-\frac{1}{2}\right)^2 + \binom{9}{8}\left(\frac{1}{2}\right)^8\left(1-\frac{1}{2}\right)^1 + \binom{9}{9}\left(\frac{1}{2}\right)^9\left(1-\frac{1}{2}\right)^0$$

   $= 0.0898.$

5. The tabulated test statistic for the sign test is simply (significance level)/2, for a two-sided H_1, and equals 0.025 for this example.
6. Reject H_0 if calculated probability is less than (significance level)/2, for a two-sided H_1. For this example, since $0.0898 > 0.025$, we do not reject H_0.
7. The median wage is not significantly different from £90 (5% level).
   Assumption The variable has a continuous distribution.

Notes

(a) Instead of P(7 or more minus signs in 9 trials for $p(-) = \frac{1}{2}$) we could have calculated P(2 or less plus signs in 9 trials for $p(+) = \frac{1}{2}$) but the answer would have been the same, because of the symmetry of the binomial when $p = 1/2$.

(b) Notice that the assumption of a continuous distribution is much less restrictive than the assumption of a normal distribution.

(c) If $n > 10$, we can alternatively use the method of Section 11.4.

(d) Since we do not use the magnitudes of the differences (between each income and £90) this test can be used even if we do not have the actual differences, but simply their signs.

11.3 SIGN TEST FOR THE MEDIAN OF A POPULATION OF DIFFERENCES, 'PAIRED' SAMPLES DATA

Example

For the example given in Section 9.10 of two methods of teaching children to read, suppose we want to decide whether the new method (N) is better than the standard method (S), but we do not wish to assume that the differences in the test scores are normally distributed. Instead we can use the sign test to decide whether the median score by the new method is greater than that by the standard method.

The difference (N score − S score) were:

$$7 \quad -2 \quad 6 \quad 4 \quad 22 \quad 15 \quad -5 \quad 1 \quad 12 \quad 15.$$

1. $H_0: p(+) = p(-) = 1/2$. Median of population of differences is zero, i.e. median of N scores = median of S scores.
2. $H_1: p(+) > p(-)$. Median of N score > median of S scores.
3. 5% significance level.

4 and 5. If the null hypothesis is true we would expect equal numbers of + and − signs. If the alternative hypothesis is true we would expect more + signs, so the null hypothesis is rejected if:

P(observed number of plus signs or more, for $p(+) = \frac{1}{2}) < 0.05$, for a one-sided H_1.

In the example we have 8 plus signs in 10 differences.

$$\begin{aligned} P(8 \text{ or more}) &= 1 - p(7 \text{ or fewer}) \\ &= 1 - 0.9453 \\ &= 0.0547, \text{ using Table D.1.} \end{aligned}$$

6. Since $0.0547 > 0.05$ (if only just), we do not reject the null hypothesis, H_0.
7. Median score by N method is not significantly greater than by the S method (5% level).
   Assumption The differences are continuous, which can be assumed for test scores even if we quote them to the nearest whole number.

Notes

(a) If $n > 10$, we can alternatively use the method of Section 11.4, which will often be quicker.
(b) This test can be used in cases where we do not know or cannot quantify the magnitudes of the differences, for example in preference testing. Only the signs of the differences are used, so we could use the convention: Brand A preferred to Brand B is recorded as a +, and so on.
(c) Differences of zero are ignored in this test.

*11.4 SIGN TEST FOR LARGE SAMPLES ($n > 10$)

The sign test for sample sizes larger than 10 is made easier by the use of a normal approximation method (similar to that used in Section 7.6) by putting:

$$\mu = \frac{n}{2} \quad \text{and} \quad \sigma = \frac{\sqrt{n}}{2}.$$

Example

Suppose that for $n = 30$ paired samples there are 20 + and 10 − differences.

1. $H_0: p(+) = p(-) = 1/2$. Median of population of differences is zero.
2. $H_1: p(+) \neq p(-)$ (two-sided).
3. 5% significance level.
4. Following the method used in the example in Section 11.2, we now need to calculate P(20 or more + signs in 30 trials, for $p(+) = \frac{1}{2}$). Instead of calculating several binomial probabilities we can put

$$\mu = \frac{n}{2} = 15, \qquad \sigma = \frac{\sqrt{n}}{2} = 2.74,$$

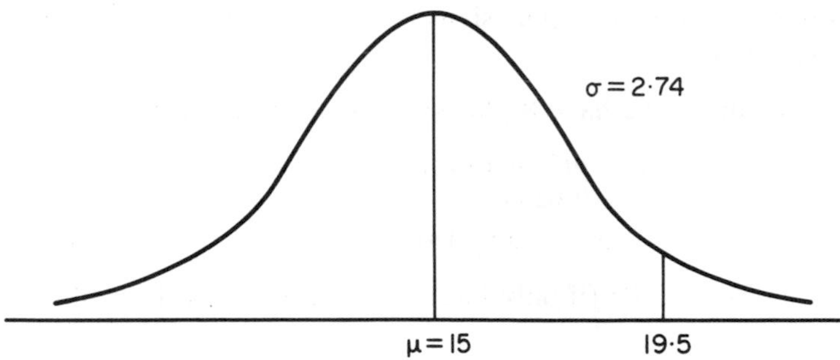

Fig. 11.1 A normal distribution, $\mu = 15$, $\sigma = 2.74$.

and use the corresponding normal distribution (see Fig. 11.1).

We also need to apply a continuity correction since '20 or more on a discrete scale' is equivalent to 'more than 19.5 on a continuous scale' (remembering that the binomial is for discrete variables, and the normal distribution is for continuous variables).

For $x = 19.5$, $Calc\ z = (x - \mu)/\sigma = (19.5 - 15)/2.74 = 1.64$.

5. $Tab\ z = 1.96$ from Table D.3(b), since this value of z corresponds to a tail-area of 0.05/2, the significance level divided by 2 since H_1 is two-sided.
6. Since $|Calc\ z| < Tab\ z$, do not reject the null hypothesis.
7. Median of differences is not significantly different from zero (5% level). *Assumption* The differences are continuous.

11.5 WILCOXON SIGNED RANK TEST FOR THE MEDIAN OF A POPULATION OF DIFFERENCES, 'PAIRED' SAMPLES DATA

In the Wilcoxon signed ranked test the hypothesis tested is the same as for the sign test. Since the former test uses the magnitudes as well as the signs of the differences it is more powerful than the sign test, and hence is a preferred method when the magnitudes are known.

The general method of obtaining the calculated test statistic for the Wilcoxon signed rank test is as follows:

> Disregarding ties (a tie is a difference of zero), the n differences are ranked without regard to sign. The sum of the ranks of the positive differences, T_+, and the sum of the ranks of the negative differences, T_-, are calculated. (A useful check is $T_+ + T_- = \frac{1}{2}n(n + 1)$.)

The smaller of T_+ and T_- is the calculated test statistic, T.

Example

Using the data of Section 11.3, the differences are as follows:

(N score − S score)	7	−2	6	4	22	15	−5	1	12	15
Ranking without regard to sign	1	−2	4	−5	6	7	12	15	15	22
The ranks are	1	2	3	4	5	6	7	$8\frac{1}{2}$	$8\frac{1}{2}$	10

Notes on tied ranks The two values in rank positions 8 and 9 are equal (to 15) and are both given the mean of the ranks they would have been given if they had differed slightly. This is an example of tied ranks.

$$T_+ = \text{sum of the ranks of the + differences}$$
$$= 1 + 3 + 5 + 6 + 7 + 8\tfrac{1}{2} + 8\tfrac{1}{2} + 10 = 49$$
$$T_- = \text{sum of the ranks of the − differences}$$
$$= 2 + 4 = 6.$$

Check $T_+ + T_- = 49 + 6 = 55$ and $\frac{1}{2}n(n+1) = 1/2 \times 10 \times 11 = 55$, so this agrees.
The smaller of T_+ and T_- is 6, so $T = 6$.

Setting out the seven-step method,

1. H_0: Median of population of differences is zero, which implies median of N scores = median of S scores.
2. H_1: Median of N scores > median of S scores.
3. 5% significance level.
4. *Calc* $T = 6$, from above.
5. *Tab* $T = 10$, from Table D.6 of Appendix D for 5% significance level, and one-sided H_1, and $n = 10$.
6. Since *Calc* $T \leqslant$ *Tab* T is true here, reject H_0.
7. Median of N scores is significantly greater than median of S scores (5% level).

   Assumption The distribution of differences is continuous and symmetrical.

Notes

(a) In step 6 we reject H_0 if *Calc* $T \leqslant$ *Tab* t, i.e. even if *Calc* T = *Tab* T.

(b) When $n > 25$, Table D.6 cannot be used. Use the method of Section 11.6 in this case.

(c) The same data have been analysed using both the sign test and the Wilcoxon test. The conclusions are not the same; using the sign test H_0 was not rejected (although the decision was a close one) and using the Wilcoxon signed rank test H_0 was rejected. Since, as we have already mentioned, the latter test is more powerful, the latter conclusion is preferred.

*11.6 WILCOXON SIGNED RANK TEST FOR LARGE SAMPLES ($n > 25$)

When $n > 25$, Table D.6 cannot be used. Instead we use a normal approximation method by putting:

$$\mu_T = \frac{n(n+1)}{4} \quad \text{and} \quad \sigma_T = \sqrt{\frac{n(n+1)(2n+1)}{24}}.$$

Example

Suppose that for $n = 30$ paired samples $T_+ = 300$, $T_- = 165$, so that $T = 165$.

1. H_0: Median of population of differences is zero.
2. H_1: Median of population of differences is not zero (two-sided).
3. 5% significance level.
4. $$\mu_T = \frac{n(n+1)}{4} = 232.5, \qquad \sigma_T = \sqrt{\frac{n(n+1)(2n+1)}{24}} = 48.6,$$

   see Fig. 11.2.

   $$Calc\ z = \frac{T - \mu_T}{\sigma_T} = \frac{165.5 - 232.5}{48.6} = 1.38.$$

   Note the use of the continuity correction as in Section 11.4.
5. *Tab* $z = 1.96$ from Table D.3(b), since this value of z corresponds to a tail area of 0.05/2, the significance level divided by 2, since H_1 is two-sided.

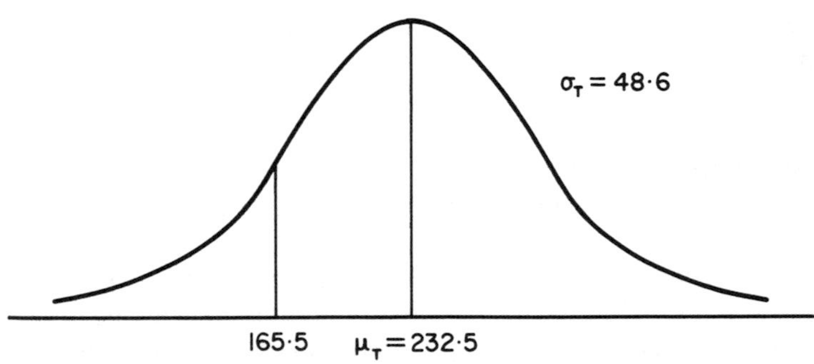

Fig. 11.2 A normal distribution, $\mu_T = 232.5$, $\sigma_T = 48.6$.

6. Since $|Calc\ z| < Tab\ z$, do not reject H_0.
7. The median of differences is not significantly different from zero (5% level).
   Assumption The distribution of differences is continuous and symmetrical.

11.7 MANN–WHITNEY *U* TEST FOR THE DIFFERENCE IN THE MEDIANS OF TWO POPULATIONS, 'UNPAIRED' SAMPLES DATA

If we cannot justify the assumptions required in the unpaired samples *t*-test (Section 10.13), the Mann–Whitney *U* test may be used to test.

H_0: The two populations have distributions which are identical in all respects.
H_1: The two populations have distributions with different medians, but are otherwise identical.

The alternative hypothesis here is two-sided, but one-sided alternatives can also be specified.

The general method of obtaining the calculated test statistic for the Mann–Whitney *U* test is as follows:

Letting n_1 and n_2 be the sizes of the samples drawn from the two populations, the $(n_1 + n_2)$ sample values are ranked as one group. The sum of the ranks of the values in each sample, R_1 and R_2, are calculated. Then U_1 and U_2 are calculated using:

$$U_1 = n_1 n_2 + \tfrac{1}{2}n_1(n_1 + 1) - R_1$$

$$U_2 = n_1 n_2 + \tfrac{1}{2}n_2(n_2 + 1) - R_2.$$

(A useful check is $U_1 + U_2 = n_1 n_2$.) The smaller of U_1 and U_2 is the calculated test statistic, *U*.

Example

As part of an investigation into factors underlying the capacity for exercise a random sample of 11 factory workers took part in an exercise test. Their heart rates in beats per minute at a given level of oxygen consumption were:

112 104 109 107 149 127 125 152 103 111 132.

A random sample of nine racing cyclists also took part in the same exercise test and their heart rates were:

91 111 115 123 83 112 115 84 120

Fig. 11.3 Heart rates of factory workers and cyclists.

These data do not look convincingly normal if plotted on a cross-diagram. We will use a Mann–Whitney U test. It is convenient to represent the data on a rough scale prior to ranking them (see Fig. 11.3).

Ranking all 20 values as one group, giving equal values the average of the ranks they would have been given if they had differed slightly, we obtain Fig. 11.4.

Factory workers ranks			$4,5,6,7,8\frac{1}{2},10\frac{1}{2}$		16,17	18	19,20
Cyclists ranks	1,2	3	$8\frac{1}{2},10\frac{1}{2},12\frac{1}{2}$, $12\frac{1}{2}$	14,15			

Fig. 11.4 Ranks of heart rates of factory workers and cyclists.

$$n_1 = 11, \quad n_2 = 9,$$

$$R_1 = 4 + 5 + 6 + 7 + 8\tfrac{1}{2} + 10\tfrac{1}{2} + 16 + 17 + 18 + 19 + 20 = 131,$$

$$R_2 = 1 + 2 + 3 + 8\tfrac{1}{2} + 10\tfrac{1}{2} + 12\tfrac{1}{2} + 12\tfrac{1}{2} + 14 + 15 = 79,$$

$$U_1 = n_1 n_2 + \tfrac{1}{2}n_1(n_1 + 1) - R_1 = 11 \times 9 + \tfrac{1}{2} \times 11 \times 12 - 131 = 34$$
$$U_2 = n_1 n_2 + \tfrac{1}{2}n_2(n_2 + 1) - R_2 = 11 \times 9 + \tfrac{1}{2} \times 9 \times 10 - 79 = 65.$$

Check $U_1 + U_2 = 34 + 65 = 99$, $n_1 n_2 = 99$, so this checks.

The smaller of U_1 and U_2 is 34, so $U = 34$.

Setting out the seven-step method,

1. H_0 : The populations of heart rates for cyclists and factory workers have identical distributions.
2. H_1 : The distributions have different medians, but are otherwise identical (two-sided).
3. 5% significance level.
4. *Calc* $U = 34$, from above.
5. Tab $U = 23$, from Table D.7 of Appendix D for 5% significance level, two-sided H_1, $n_1 = 11$, $n_2 = 9$.
6. Since *Calc* $U >$ *Tab* U, do not reject H_0.
7. The median of heart rates for factory workers and cyclists are not significantly different (5% level).

   Assumption The variable is continuous. Since the number of beats/minute is large and may be an average of several values, this assumption is reasonable.

Notes

(a) In step 6 we reject H_0 if *Calc* $U \leqslant$ *Tab* U, i.e. even if *Calc* $U =$ *Tab* U.

(b) When n_1 or n_2 is greater than 20, Table D.7 cannot be used. Use the method of Section 11.8 instead.

*11.8 MANN–WHITNEY *U* TEST FOR LARGE SAMPLES ($n_1 > 20, n_2 > 20$)

When n_1, $n_2 > 20$, we use a normal approximation method by putting:

$$\mu_U = \frac{n_1 n_2}{2} \quad \text{and} \quad \sigma_U = \sqrt{\frac{n_1 n_2(n_1 + n_2 + 1)}{12}}.$$

Example

Suppose that for two unpaired samples of size $n_1 = 25$, $n_2 = 30$, we obtain $R_1 = 575$, $R_2 = 965$, $U_1 = 500$, $U_2 = 250$, so $U = 250$.

1. H_0: The two populations have identical distributions.
2. H_1: The two populations have distributions with different medians, but are otherwise identical (two-sided).
3. 5% significance level.
4.

$$\mu_U = \frac{n_1 n_2}{2} = 375, \qquad \sigma_U = \sqrt{\frac{n_1 n_2(n_1 + n_2 + 1)}{12}} = 59.2,$$

see Fig. 11.5.

$$\textit{Calc } z = \frac{U - \mu_U}{\sigma_U} = \frac{250.5 - 375}{59.2} = -2.10.$$

Note the use of the continuity correction as in Section 11.4.

5. *Tab* $z = 1.96$ from Table D.3(b), since this value of z corresponds to a

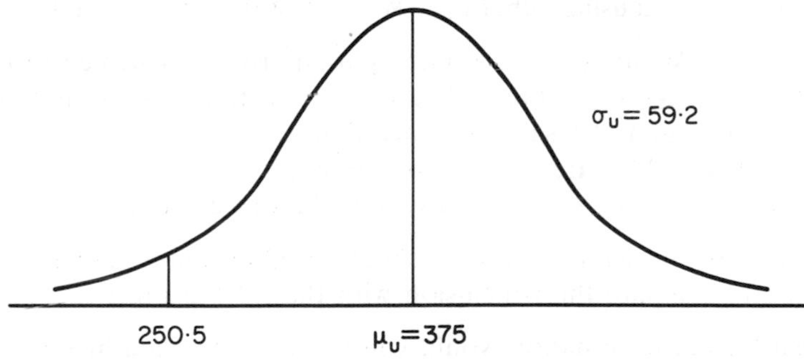

Fig. 11.5 A normal distribution, $\mu_U = 375$, $\sigma_U = 59.2$.

tail area of 0.05/2, the significance level divided by 2, since H_1 is two-sided.

6. Since $|Calc\ z| > Tab\ z$, reject H_0.
7. The medians are significantly different (5% level).
Assumption The variable is continuous.

11.9 SUMMARY

Three non-parametric tests, the sign test, the Wilcoxon signed rank test and the Mann–Whitney U test are described for the small- and large-sample cases. These tests make fewer assumptions than the corresponding t tests, but are less powerful. Table 11.2 summarizes the various tests covered in this chapter.

WORKSHEET 11: SIGN TEST, WILCOXON SIGNED RANK TEST, MANN–WHITNEY U TEST

Fill in the gaps in Questions 1, 2, 3 and 5.

1. Non-parametric tests are used to test ——— in cases where the ——— of the corresponding parametric tests are not valid.
2. However, when the ——— are valid it is better to use parametric tests because they are more ——— than the corresponding non-parametric tests.
3. Power is the risk of rejecting the ——— hypothesis when the ——— hypothesis is correct. The ——— the power of an hypothesis test, the better.
4. The sign test and the Wilcoxon signed rank test may both be used on paired samples data. Give examples of data which could
   (a) only be analysed using the sign test,
   (b) be analysed using either test. Which test is preferable in this case?
5. The Mann–Whitney U test is a non-parametric test which corresponds to the ——— ——— t-test. The latter is a more ——— test if two ——— are valid. These ——— are that:
   (a) both variables are ——— distributed;
   (b) the ——— ——— of the two populations are equal.
6. Re-analyse the data from Worksheet 10, Question 7, using the sign test, and compare the conclusion with that of the t-test.
7. What further information would you need for the data in Worksheet 10, Question 8, in order to carry out a sign test?

Table 11.2 Non-parametric hypothesis tests

Test	*Parameter*	*Case*	*Assumption*	*Reject* H_0 *if*[†]
Sign	Median of a population	$n \leqslant 10$	Variable has a continuous distribution	*Calc binomial probability* $\leqslant \frac{\text{significance level}}{2}$
Sign	Median of a population of differences	$n \leqslant 10$	Variable has a continuous distribution	*Calc binomial probability* $\leqslant \frac{\text{significance level}}{2}$
Sign	Median of a population of differences	$n > 10$	Variable has a continuous distribution	$\lvert Calc\ z \rvert > Tab\ z$
Wilcoxon	Median of a population of differences	$n \leqslant 25$	Variable has a continuous and symmetrical distribution	$Calc\ T \leqslant Tab\ T$
Wilcoxon	Median of a population of differences	$n > 25$	Variable has a continuous and symmetrical distribution	$\lvert Calc\ z \rvert > Tab\ z$
Mann–Whitney	Difference between the medians of two populations	$n_1, n_2 \leqslant 20$	Variable has a continuous distribution	$Calc\ U \leqslant Tab\ U$
Mann–Whitney	Difference between the medians of two populations	$n_1, n_2 > 20$	Variable has a continuous distribution	$\lvert Calc\ z \rvert > Tab\ z$

[†] The 'decision rules' in this column are for two-sided alternative hypotheses only, except for T and U.

8. Re-analyse the data from Worksheet 10, Question 14, using the Wilcoxon signed rank test.

9. A psychologist tested eight subjects, randomly chosen from the eleven-year-old boys taught in comprehensive schools in a city, using a standard aptitude test. The scores were:

135 103 129 96 122 140 110 91

(a) Later the same subjects received a new improved aptitude test and the scores (in the same order of subject) were:

125 102 117 94 120 130 110 92

Is there a significant difference between the average scores for the two tests? Use an appropriate non-parametric test.

(b) Now assume that the scores for the second test refer to an independent second random sample of eight subjects. Is there a significant difference between the average scores for the two tests? Again use an appropriate non-parametric test.

10. An investigation was carried out at a trout farm to find the effect of a new feeding compound. Twenty fry (newly born fish) were randomly divided in two equal groups. Both groups were then kept under the same environmental conditions, but one group was fed with a standard feeding compound and the other group was fed with the new feeding compound. After a given period the fish were weighed:

Weight of fry (g)	
Standard compound	*New compound*
510	521
507	476
490	489
496	512
523	521
508	498
534	505
497	547
491	542
506	492

Analyse these data using a non-parametric test.

11. Two brands of car tyre were tested in simulated road trials. The 'distances' travelled by 12 tyres of one brand and 12 tyres of the other

brand before their treads had worn below the legal minimum limit was recorded to the nearest thousand kilometres:

Brand 1	47	44	39	41	39	42	51	44	55	45	49	46
Brand 2	43	33	40	38	31	39	34	40	35	37	38	32

Is one brand better than the other? Use a non-parametric test.

12

Association of categorical variables

12.1 INTRODUCTION

The inferential methods we discussed in Chapters 9, 10 and 11 involved data for one variable measured on a number of 'individuals' where the variable was numerical (either continuous, discrete, or ranked). We now turn to data for categorical variables (refer to Section 1.2). Also, instead of one variable (univariate) data we will discuss two variable (bivariate) data.

So we will be concerned in this chapter with investigations in which two categorical variables are recorded for a number of 'individuals'. Such data may be set out in two-way contingency tables (refer to Section 3.5). We will decide whether the variables are independent or whether they are associated by performing a hypothesis test, the χ^2 (chi-squared) test for independence.

12.2 CONTINGENCY TABLES

Remember that a categorical variable is one which is not quantitative, but can take 'values' which are non-numerical categories or classes. So for an individual person, sex, religion and employment status are three examples of categorical variables. Other examples are hair colour and temperament. Suppose we record the hair colour and temperament of a number of individuals as follows:

Hair colour	*Temperament*
Black	Easy-going
Brown	Lively
Blond	Friendly
Red	Fiery
⋮	⋮

Assuming for each categorical variable that there is a finite number of mutually exclusive (non-overlapping) and exhaustive categories these data could be represented in a two-way table called a contingency table (see Table 12.1).

Table 12.1 Contingency table for hair colour and temperament

Hair colour	Temperament				
	Easy-going	Lively	Friendly	Fiery	• • •
Black					
Brown					
Blond					
Red					
⋮					

The entries in the various 'cells' of the table would be the number of individuals, or frequencies, for the various cross-categories. For example, if we have five rows and six columns there would be $5 \times 6 = 30$ cells, and 30 frequencies. In general the '$r \times c$' contingency table has r rows, c columns, $r \times c$ cells and $r \times c$ frequencies.

Suppose we want to decide whether the two categorical variables are associated or independent. We will restrict ourselves initially to a 2×2 contingency table by considering an example in which hair colour is designated as red, or not red, and temperament is designated as 'good temper' and 'bad temper'. Then the data for a random sample of 100 individuals might be as in Table 12.2.

Table 12.2 Contingency table for hair colour and temperament of 100 individuals

	Temper	
Hair colour	*Bad*	*Good*
Red	30	10
Not Red	20	40

12.3 χ^2 TEST FOR INDEPENDENCE, 2×2 CONTINGENCY TABLE DATA

We will carry out the usual seven-step method of hypothesis testing.

1. H_0: The variables, hair colour and temper are independent, i.e. there is no association between the variables.

2. H_1 : The variables are not independent, they are associated (two-sided).
3. 5% significance level.
4. Calculated test statistic. If all expected frequencies, E, are ≥ 5 the test statistic is

$$Calc\ \chi^2 = \Sigma \frac{(O - E)^2}{E}$$

for the general $r \times c$ table, but we use

$$Calc\ \chi^2 = \Sigma \frac{(|O - E| - \frac{1}{2})^2}{E}$$

for the 2×2 table. The latter formula includes what is called Yates's continuity correction.

Here O stands for observed frequency. The values in Table 12.2 are the observed frequencies. E stands for expected frequency. The expected frequencies are those we would expect if we assume (for the purposes of calculating them) that the null hypothesis is true and we keep the row and column totals (as in Table 12.3 below) fixed. We calculate the E values for each cell in the table using the formula:

$$E = \frac{\text{row total} \times \text{column total}}{\text{grand total}}.$$

It is convenient to put the E values in brackets next to the O values (see Table 12.3).

Example of the calculation of the E value for row 2, column 1.

$$E = \frac{\text{row 2 total} \times \text{col 1 total}}{\text{grand total}} = \frac{60 \times 50}{100} = 30.$$

Table 12.3 Observed (and expected) frequencies

	Temper		
Hair colour	*Bad*	*Good*	*Row total*
Red	30(20)	10(20)	40
Not red	20(30)	40(30)	60
Column total	50	50	*Grand total* 100

The symbol Σ in the formula for *Calc* χ^2 means that we sum over all cells in the table. The term $|O - E|$ in the formula for χ^2 for a 2×2 table means that we take the magnitude of the $O - E$ values and ignore the sign (see calculation below). For the data in Table 12.3,

$$Calc\ \chi^2 = \frac{(|30-20|-\frac{1}{2})^2}{20} + \frac{(|10-20|-\frac{1}{2})^2}{20} + \frac{(|20-30|-\frac{1}{2})^2}{30} + \frac{(|40-30|-\frac{1}{2})^2}{30}$$

$$= \frac{(10-\frac{1}{2})^2}{20} + \frac{(10-\frac{1}{2})^2}{20} + \frac{(10-\frac{1}{2})^2}{30} + \frac{(10-\frac{1}{2})^2}{30}$$

$$= 15.04.$$

5. *Tabulated test statistic*

   Tab χ^2 is obtained from Table D.8 of Appendix D, and we enter the tables for α = significance level, even though H_1 is two-sided, and $\nu = (r-1)(c-1)$. For the example of a 5% significance level and a 2×2 table, *Tab* $\chi^2 = 3.84$, for $\alpha = 0.05$, $\nu = (2-1)(2-1) = 1$.
6. If *Calc* $\chi^2 >$ *Tab* χ^2, reject the null hypothesis. For the example *Calc* $\chi^2 >$ *Tab* χ^2, so the null hypothesis is rejected.
7. We conclude that there is significant association between hair colour and temperament (5% level).

   Assumption The observations must be independent; the best way of ensuring this is by random sampling. An example of dependent data would be if the data represented in Table 12.3 were obtained from 50 pairs of twins.

Note

You should read the next section before trying Worksheet 12.

12.4 FURTHER NOTES ON THE χ^2 TEST FOR INDEPENDENCE, CONTINGENCY TABLE DATA

(a) The null hypothesis of independence may also be expressed in terms of proportions (or probabilities). So for the data in Table 12.3, we could write:

$$H_0: \text{P(bad temper|red hair)} = \text{P(bad temper|not red hair)}$$

(refer to Section 5.9 on conditional probabilities, if necessary).

(b) A two-sided alternative is taken because the calculation of χ^2 requires the differences, $(O - E)$, to be squared, so that positive and negative differences of the same magnitude have the same effect on the calculation. If the null hypothesis is rejected we can see the direction of the association by going back to the table containing O and E values. For the example in Table 12.3 H_0 was rejected. We note that the O value is greater than the E value for the cell Red/Bad. Since the E value is calculated assuming independence, these data support the view that red hair and bad temper are associated.

(c) The formula for *Calc* χ^2 is theoretically valid only if all the E values are sufficiently large, and $E \geqslant 5$ is the accepted condition to apply (the O

values are not restricted in this way). If any E value is <5, it may be sensible to combine adjacent rows (or columns) to form a smaller contingency table with all E values $\geqslant 5$. For a 2×2 table this is clearly not an option. Instead a different test may be applied, called the Fisher exact test, which we will not discuss in this book. (Refer, for example, to Bailey (1969) listed in Appendix E).

(d) The fact that we may conclude that there is a significant association between the variables does not necessarily imply cause and effect (we will make a similar statement in connection with significant correlation coefficients in Chapter 13). So we cannot conclude for the example that rejection of H_0 implies that red hair causes an individual to be bad-tempered or that having a bad temper causes an individual's hair to turn red! Assuming there is really an association (and the data in Table 12.3 are fictitious) there are probably other factors which are as responsible for the association, such as heredity.

(e) The formula for the degrees of freedom for the χ^2 test can be justified in terms of the definition of Section 9.7. For a 2×2 table, once we have calculated one E value, the other three are determined by the restriction that the row and column totals are fixed (the sum of the E values in any row, or column, is the same as the sum of the O values). So for a 2×2 table, there is only one degree of freedom, and a similar argument can be applied to justify the use of $(r - 1)(c - 1)$ degrees of freedom for the general $r \times c$ table.

12.5 χ^2 TEST FOR INDEPENDENCE, 3×3 TABLE

Because so many points have been discussed in connection with the χ^2 test for independence, a further example will now be given, this time for a 3×3 table.

Example

Sixty workers were randomly selected and asked to give their opinion on a new pension scheme which their employer was considering. Of 10 workers with a high income, 8 were in favour of the new scheme, 1 was undecided and 1 was against the new scheme. Of 25 workers with an average income, the numbers in favour, undecided and against were 7, 3 and 15 respectively. Of 25 workers with a low income the corresponding numbers were 2, 10 and 13 respectively. Are the opinions of workers independent of their income? (Although income is quantitative it is treated as categorical here for the purposes of this example).

Table 12.4 can be formed from the above information. E values (calculated as in Section 12.3) are given in brackets to one decimal place (which is sufficient).

Table 12.4 3 × 3 contingency table for income and opinion of 60 workers

	Opinion on a new pension scheme			
Income	*In favour*	*Undecided*	*Against*	*Row total*
High	8(2.8)	1(2.3)	1(4.8)	10
Average	7(7.1)	3(5.8)	15(12.1)	25
Low	2(7.1)	10(5.8)	13(12.1)	25
Column total	17	14	29	60

Since we have three E values less than 5 in this table we cannot apply the χ^2 test to these data as they stand. We can, however, combine the top two rows as in Table 12.5.

Table 12.5 2 × 3 contingency table for income and opinion of 60 workers

	Opinion on a new pension scheme			
Income	*In favour*	*Undecided*	*Against*	*Row total*
High or Average	15(9.9)	4(8.2)	16(16.9)	35
Low	2(7.1)	10(5.8)	13(12.1)	25
Column total	17	14	29	60

All E values are now greater than 5, so the seven-step method can now be applied:

1. H_0: Opinion (on a new pension scheme) and income are independent.
2. H_1: Opinion and income are associated (two-sided).
3. 5% significance level.
4.

$$\textit{Calc}\ \chi^2 = \Sigma\frac{(O-E)^2}{E}, \text{ for this } 2 \times 3 \text{ table}$$

$$= \frac{(15-9.9)^2}{9.9} + \frac{(4-8.2)^2}{8.2} + \frac{(16-16.9)^2}{16.9}$$

$$+ \frac{(2-7.1)^2}{7.1} + \frac{(10-5.8)^2}{5.8} + \frac{(13-12.1)^2}{12.1}$$

$$= 11.6.$$

5. Tab $\chi^2 = 5.99$ for $\alpha = 0.05$, $\nu = (2 - 1)(3 - 1) = 2$.
6. Since *Calc* χ^2 > *Tab* χ^2, reject the null hypothesis.
7. Opinion and income are significantly associated. Looking at the cells with the greatest contribution to the *Calc* χ^2 we conclude that
   (a) more of the high or average income workers are in favour; and
   (b) more of the low income workers are undecided, than we would expect if income and opinion were independent.

12.6 SUMMARY

Inferences from bivariate categorical data were discussed. Such data, collected from a random sample of 'individuals', may be displayed as observed frequencies (O) in two-way contingency tables. The null hypothesis of independence between the variables is tested using the χ^2 statistic where:

$$Calc\ \chi^2 = \Sigma\frac{(O - E)^2}{E}$$

if the degrees of freedom, $(r - 1)(c - 1)$, are greater than 1 or

$$Calc\ \chi^2 = \Sigma\frac{(|O - E| - \frac{1}{2})^2}{E}$$

if the degrees of freedom equal 1.

The expected frequencies (E) are calculated using:

$$E = \frac{\text{row total} \times \text{column total}}{\text{grand total}}.$$

If any E value is less than 5, the formula for *Calc* χ^2 is invalid, and alternative methods must be considered. Rejection of the null hypothesis of independence does not imply cause and effect.

WORKSHEET 12: ASSOCIATION OF CATEGORICAL VARIABLES

Fill in the gaps in Questions 1–7.

1. A categorical variable can only take 'values' which are non ———.
2. If we collect data for two categorical variables for a number of individuals, the data may be displayed in a two-way or ——— table. In such a table the numbers in the various cells of the table are the number of ——— in each cross-category and are referred to as ——— frequencies.
3. The null hypothesis in the analysis of contingency table data is that the two categorical variables are ———.

4. In order to calculate the χ^2 test statistic we first calculate the ——— frequencies, using the formula $E =$ ———————.

5. If all the E values are greater than or equal to ———, the test statistic *Calc* χ^2 = ——— is calculated. Since the E values are calculated assuming the null hypothesis is true, high values of *Calc* χ^2 will tend to lead to the ——— of the null hypothesis.

6. The degrees of freedom for *Tab* χ^2 are (———)(———) for a contingency table with r rows and c columns, so for a 2×2 contingency table there are ——— degrees of freedom.

7. For a 2×2 contingency table we reject the null hypothesis if *Calc* $\chi^2 >$ ——, for a 5% significance level.

8. Of 40 rented television sets the tubes of 9 sets burnt out within the guarantee period of 2 years. Of 60 bought sets, the tubes of 5 sets burnt out within 2 years. Test the hypothesis that the proportion of burnt sets is independent of whether they were bought or rented, assuming that the 100 sets referred to are a random sample of all sets.

9. For four garages in a city selling the same brand of four-star petrol the following are the numbers of male and female car drivers calling for petrol between 5 pm and 6 pm on a given day. Is there any evidence that the proportion of male to female varies from one garage to another?

Sex of driver	*Garages*				
	A	*B*	*C*	*D*	*Totals*
Male	25	50	20	25	120
Female	10	50	5	15	80
Totals	35	100	25	40	200

10. The examination results of 50 students, and their attendance (%) on a course, were as follows:

Attendance	*Exam result*		
	Pass	*Fail*	*Totals*
Over 70%	20	5	25
30%–70%	10	5	15
Under 30%	5	5	10
Totals	35	15	50

Is good attendance associated with a higher chance of passing the examination?

11. Two types of sandstone were investigated for the presence of three types of mollusc. The number of occurrences were:

	Type of mollusc		
	A	*B*	*C*
Sandstone 1	15	30	12
Sandstone 2	15	9	6

Is there enough evidence to suggest that the proportion of the three types of mollusc is different for the two types of sandstone?

12. In a survey of pig farms it is suspected that the occurrence of a particular disease may be associated with the method of feeding. Methods of feeding are grouped into two categories, A and B. Of 27 farms on which the disease occurred, 16 used method A, and of 173 farms on which the disease had not occurred, 84 used method A. Test for independence between the method of feeding and occurrence of the disease.

13. A random sample of 300 people took part in a market research survey to find their reaction to the flavour of a new toothpaste. The results were as follows:

	Men	*Women*	*Children*
Like flavour	20	30	100
Dislike flavour	50	40	30
No opinion	10	20	0

What conclusion can be drawn from these data, comparing the likes and dislikes of:

(a) Men, women and children?

(b) Men and women?

Correlation of quantitative variables

Besides, in many instances it is impossible to determine whether these are causes or effects.

13.1 INTRODUCTION

In the previous chapter we discussed tests of the association of two categorical variables. If, instead, we are interested in the association of two quantitative (numerical) variables, we may:

(a) Summarize the sample data graphically in a scatter diagram (refer to Fig. 3.7).
(b) Calculate a numerical measure of the degree of association called a correlation coefficient.
(c) Carry out an hypothesis test of independence between the variables and, if we conclude that there is significant correlation between the variables, interpret this conclusion with great care!

We will discuss two correlation coefficients:

(1) Pearson's (product moment) correlation coefficient, which we use if we can reasonably assume that the variables are both normally distributed.
(2) Spearman's rank correlation coefficient, which we use if we cannot reasonably assume normality.

If the variables are normally distributed the hypothesis test for Pearson's coefficient is more powerful than the test for Spearman's coefficient.

13.2 PEARSON'S CORRELATION COEFFICIENT

Example

Suppose we record the heights and weights of a random sample of six adults (see Table 13.1). It is reasonable to assume that these variables are normally distributed, so the Pearson correlation coefficient is the appropriate measure of the degree of association between height and weight.

Table 13.1 Heights and weights of a random sample of six adults

Height (cm)	*Weight* (kg)
170	57
175	64
176	70
178	76
183	71
185	82

A scatter diagram for these data can be drawn (see Fig. 13.1).

We will discuss this scatter diagram later once we have calculated the value of Pearson's correlation coefficient – we use the symbol r to represent the sample value of this coefficient, and ϱ (rho) to represent the population value. The formula for r is:

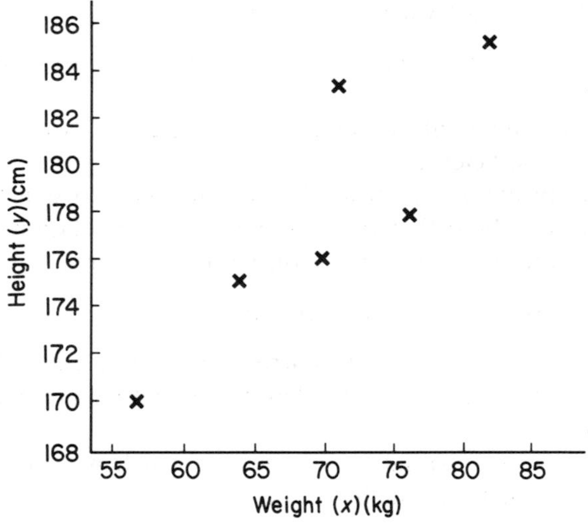

Fig. 13.1 Scatter diagram for heights and weights of a random sample of six adults.

$$r = \frac{\Sigma xy - \dfrac{\Sigma x \Sigma y}{n}}{\sqrt{\left\{\left[\Sigma x^2 - \dfrac{(\Sigma x)^2}{n}\right]\left[\Sigma y^2 - \dfrac{(\Sigma y)^2}{n}\right]\right\}}}$$

where one of our variables is the x variable, the other is the y variable, and n is the number of individuals. In correlation it is an arbitrary decision as to which variable we call x and which we call y. Suppose we decide as in the scatter diagram, Fig. 13.1, then in the formula for r, Σx means the sum of the weights, etc.

For the data in Table 13.1.

$$\begin{aligned}
\Sigma x &= 57 + 64 + 70 + 76 + 71 + 82 &&= 420 \\
\Sigma x^2 &= 57^2 + 64^2 + \cdots + 82^2 &&= 29\,786 \\
\Sigma y &= 170 + 175 + \cdots + 185 &&= 1\,067 \\
\Sigma y^2 &= 170^2 + 175^2 + \cdots + 185^2 &&= 189\,899 \\
\Sigma xy &= (57 \times 170) + (64 \times 175) + \cdots + (82 \times 185) &&= 74\,901 \\
n &= 6
\end{aligned}$$

$$\begin{aligned}
r &= \frac{74\,901 - \dfrac{420 \times 1\,067}{6}}{\sqrt{\left\{\left[29\,786 - \dfrac{420^2}{6}\right]\left[189\,899 - \dfrac{1067^2}{6}\right]\right\}}} \\
&= \frac{211}{\sqrt{\{[386][150.8]\}}} \\
&= 0.875.
\end{aligned}$$

Check through this calculation if it is unfamiliar. What does a sample correlation coefficient of 0.875 tell us?

To put this value in perspective we can now look at the scatter diagram of Fig. 13.1, where the general impression is of increasing weight to be associated with increasing height, and vice versa. We can imagine a cigar-shaped outline round the data to emphasize this impression.

In general, if points on a scatter diagram show the same tendency (that is, as one variable increases, so does the other) and in addition the points lie on a straight line, then $r = 1$. If there is the opposite tendency (that is, as one variable increases, the other decreases) and in addition the points lie on a straight line, then $r = -1$. If there is no such tendency and the points look like the distribution of cherries in a perfectly made cherry-cake, then the value r is close to 0. These three cases are shown in Fig. 13.2.

Within the range of possible values for r from -1 to $+1$ we may describe a value of 0.875, which we calculated for the height/weight data

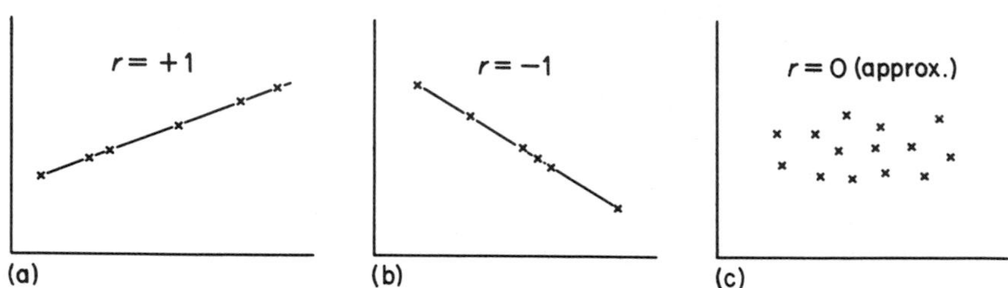

Fig. 13.2 Scatter diagrams for (a) $r = -1$; (c) $r = 0$ (approx.).

as 'high positive correlation'. But be warned. Do not judge the association between two variables simply from the value of the correlation coefficient! We must also take into account n, the number of 'individuals' contributing the sample data. Intuitively, $r = 0.875$ based on a sample of $n = 6$ individuals is not as impressive as $r = 0.875$ based on a sample of $n = 60$ individuals. Had we obtained the latter we would have more evidence of the degree of association in the population. This intuitive argument is formalized in a hypothesis test for the population correlation coefficient in the next section.

13.3 HYPOTHESIS TEST FOR PEARSON'S POPULATION CORRELATION COEFFICIENT, ϱ

Example

We will use the data and calculations of the previous section, and set out the seven-step method:

1. $H_0 : \varrho = 0$. This implies no correlation between the variables in the population.
2. $H_1 : \varrho > 0$. This implies that there is positive correlation in the population. Increasing height is associated with increasing weight.
3. 5% significance level.
4. The calculated test statistic is:

$$Calc\ t = r\sqrt{\frac{n-2}{1-r^2}}$$

(Notice that this formula includes both the values of r and n.)

$$= 0.875\sqrt{\frac{6-2}{1-0.875^2}} \quad \text{for the height/weight data,}$$

$$= 3.61.$$

5. *Tab* $t = 2.132$ from Table D.5, for

$\alpha = 0.05$ (one-sided H_1)
$\nu = (n - 2)$ (for this test)
$= 4.$

(The way to remember the number of degrees of freedom is that $n - 2$ occurs in the formula for *Calc t*.)

6. Since *Calc t* > *Tab t*, reject H_0 (refer to Section 10.10 if necessary).
7. There is significant positive correlation between height and weight. *Assumption* Height and weight are normally distributed.

Note
You should read the next section before trying Worksheet 13.

13.4 THE INTERPRETATION OF SIGNIFICANT AND NON-SIGNIFICANT CORRELATION COEFFICIENTS

The following points should be considered whenever we try to interpret correlation coefficients:

(a) A significant value of r (i.e. when the null hypothesis $H_0: \varrho = 0$ is rejected) does not necessarily imply cause and effect. For the height/weight data it clearly makes little sense to talk about 'height causing weight' or vice versa, but it might be reasonable to suggest that both variables are caused by (meaning 'depend on') a number of variables such as age, sex, heredity, diet, exercise and so on.

For the kind of examples quoted regularly by the media we must be equally vigilant. Claims like: 'eating animal fats causes heart disease', 'wearing a certain brand of perfume causes a person to be more sexually attractive', 'reducing inflation causes a reduction in unemployment' may or may not be true. They are virtually impossible to substantiate without controlling or allowing for the many other factors which may influence the chances of getting heart disease, the level of sexual attraction and the level of unemployment respectively. Such careful research is difficult, expensive and time-consuming, even in cases where the other factors may be controlled or allowed for. Where they may not be, it is misleading to draw confident conclusions.

(b) The correlation coefficient measures the linear association between the variables. So a scatter diagram may indicate non-linear correlation but the Pearson correlation coefficient may have a value close to zero. For example a random sample of 10 runners taking part in a local 'fun-run' of 10 miles may give rise to a scatter diagram like Fig. 13.3, if the age of the runner is plotted against the time to complete the course. A clear curvilinear relationship exists but the value of r would be close to zero.

(c) A few outlying points may have disproportionate effect on the value of

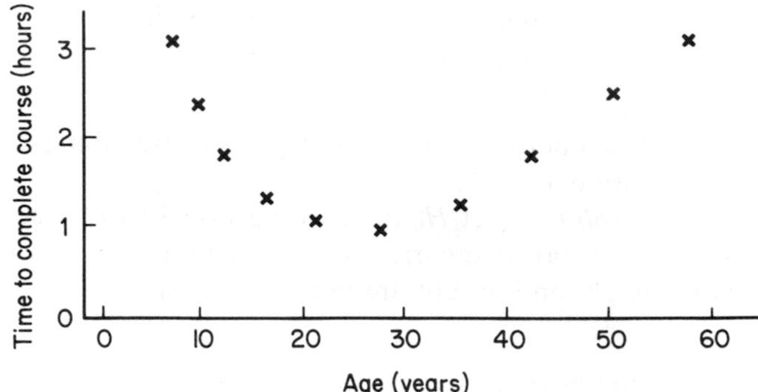

Fig. 13.3 Scatter diagram for age of runner and time to complete course for a random sample of 10 runners.

r, as in Fig. 13.4. In the left-hand diagram the inclusion of the outlier would give a smaller value of r than if it were excluded from the calculation. In the right-hand diagram the inclusion of the outlier would give a larger value of r.

In fact, in both cases the assumption that both variables are normally distributed looks suspect. In the left-hand diagram the outlier has a value of y which is far away from the other values, and in the right-hand diagram both the x and y values of the outlier are extreme. We must not discard outliers simply because they do not seem to fit into the pattern of the other points. We are justified in suspecting that some mistake may have been made in calculating the x and y values, or in plotting the point, or in some other way.

(d) The value of r may be restricted, and be non-significant in an hypothesis test, because the ranges of the x and y variables are restricted. For example, the value of r between students' A-level count and their performance in a particular college of higher education may

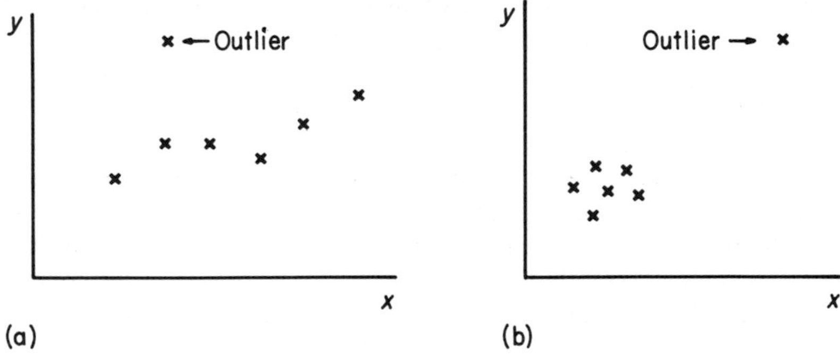

Fig. 13.4 Two scatter diagrams with an 'outlier'.

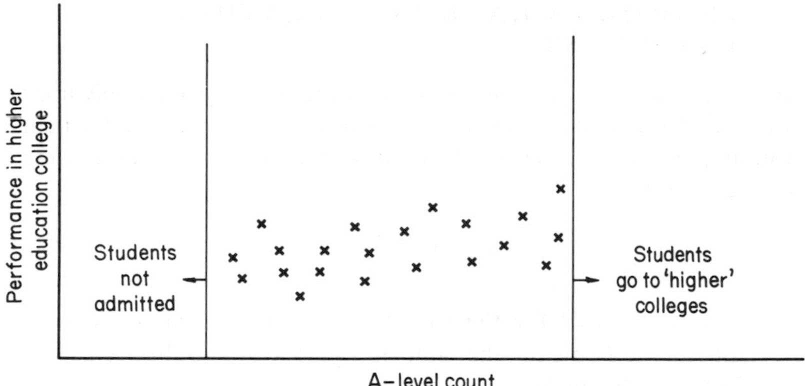

Fig. 13.5 Scatter diagram for performance in higher education college and A-level count.

be restricted by the fact that, for that college, only students with A-level counts above a certain minimum are admitted while students with A-level counts which are very high may choose to go to other colleges (higher up the 'pecking order'). The value of r for students actually admitted to the particular college may be lower than if entry was unrestricted (see Fig. 13.5).

(e) Nonsense correlations may result if common sense indicates that two variables, which are clearly unconnected, have increased or decreased in step over a period of time. There are many examples of this type of correlation: the number of violent crimes committed and doctors' salaries have increased over the last 10 years, and the correlation coefficient (for the 10 'individual' years) may be significant. Clearly it would be nonsense to conclude that giving more money to doctors increases the incidence of violent crime. Another nice example is the observation that in a Swedish town, it was observed that in years when relatively more storks built their nests on house-chimneys, relatively more babies were born in the town, and vice versa. A scatter diagram would have shown a possibly significant positive correlation, hence the idea that storks bring babies was seemingly supported.

It may occur to you that, with all the above reservations, there is little to be gained by calculating the value of a correlation coefficient and testing it for significance. The interpretation we can place on a significant value of r is that 'such a value is unlikely to have arisen by chance if there is no correlation in the population, so it is reasonable to conclude that there is some correlation in the population'. To extend this conclusion to one of cause and effect, for example, requires much more information about other possible causal variables and consideration of the points made above in this section.

13.5 SPEARMAN'S RANK CORRELATION COEFFICIENT

If the two quantitative variables of interest are not approximately normally distributed, Spearman's rank correlation coefficient may be calculated by ranking the sample data separately for each variable and using the following formula:

$$r_s = 1 - \frac{6\Sigma d^2}{n^3 - n}$$

where r_s is the symbol for the sample value of Spearman's correlation coefficient, and Σd^2 means the sum of squares of the differences in the ranks of the n individuals.

A non-parametric hypothesis test may then be carried out.

Example

A random sample of ten students was asked to rate two courses they had all taken on a ten-point scale. A rating of 1 means 'absolutely dreadful', a rating of 10 means 'absolutely wonderful' (see Table 13.2).

Table 13.2 Two courses rated by 10 students

Statistics rating	*Mathematics rating*
7	6
6	6
3	5
8	7
2	5
6	3
7	9
7	4
10	7
4	8

Here we are not interested in whether one course has a higher mean rating than the other (in which case a Wilcoxon signed rank test would be appropriate, although unnecessary here since the sample mean ratings are equal) but we are interested in whether there is a significant correlation between the ratings. In other words, do students who rate one course highly, relative to the ratings of other students, also tend to rate the other course highly, relative to the ratings of other students, and vice versa? The scatter diagram of Fig. 13.6 indicates the correlation may be positive but small, and we will hopefully confirm this subjective judgement when we calculate the sample value of the Spearman rank correlation coefficient (see Table 13.3 and what follows).

Table 13.3 The ranks and squared differences in ranks (d) for data in Table 13.2.

Statistics rating	*Mathematics rating*	*Ranks of statistics rating*	*Ranks of mathematics rating*	d^2
7	6	7	$5\frac{1}{2}$	2.25
6	6	$4\frac{1}{2}$	$5\frac{1}{2}$	1.00
3	5	2	$3\frac{1}{2}$	2.25
8	7	9	$7\frac{1}{2}$	2.25
2	5	1	$3\frac{1}{2}$	6.25
6	3	$4\frac{1}{2}$	1	12.25
7	9	7	10	9.00
7	4	7	2	25.00
10	7	10	$7\frac{1}{2}$	6.25
4	8	3	9	36.00
			Σd^2 =	102.50

Note on tied ranks

When two (or more) ratings for a course are equal they are given the average of the ranks they would have been given if they had differed slightly.

From Table 13.3, we now calculate r_s:

$$r_s = 1 - \frac{6\Sigma d^2}{n^3 - n}$$

$$= 1 - \frac{6 \times 102.50}{10^3 - 10}$$

$$= 0.379.$$

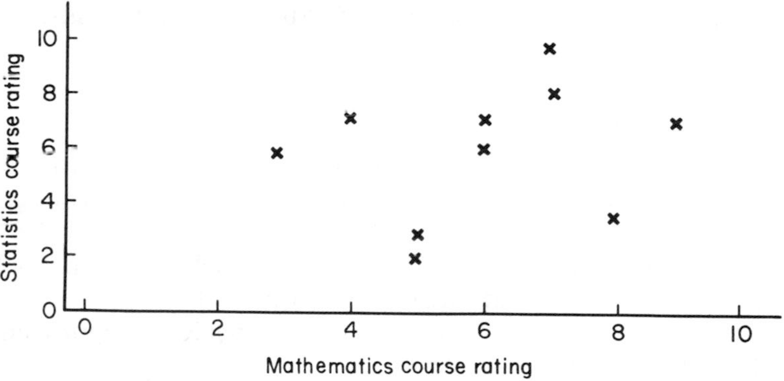

Fig. 13.6 Scatter diagram for ratings of 10 students taking courses in statistics and mathematics.

What does a sample value of 0.379 for r_s tell us? The possible range of values for r_s is -1 to $+1$ (the same as the possible range for Pearson's r).

If $r_s = +1$ there is perfect agreement in the rankings. If $r_s = -1$ there is perfect disagreement (the highest rank for one variable corresponding to the lowest rank for the other variable, and so on). If $r_s = 0$, a particular rank for one variable may correspond with any rank for the other variable. So a value of $r_s = 0.379$ is low positive correlation (as we suspected from the scatter diagram), but this value should not be judged in isolation, since (as with Pearson's r) we must also take into account the number of 'individuals' n, which we do by carrying out an hypothesis test (see next section).

13.6 HYPOTHESIS TEST FOR SPEARMAN'S RANK CORRELATION COEFFICIENT

Example

Using the sample data of the example in the previous section:

1. H_0: The rankings in statistics and mathematics are uncorrelated.
2. H_1: High ranks in statistics either correspond to high ranks in mathematics or to low ranks in mathematics (two-sided alternative).
3. 5% significance level.
4. *Calc* $r_s = 0.379$, from the previous section.
5. *Tab* $r_s = 0.648$, from Table D.9 of Appendix D for $n = 10$, two-sided H_1 and 5% significance level.
6. Since $|Calc\ r_s| < Tab\ r_s$, do not reject H_0.
7. There is no significant correlation between the rankings in mathematics and statistics.

   Assumption We must be able to rank each variable.

The notes in Section 13.4 on the interpretation of correlation coefficients apply equally to both the Pearson and the Spearman coefficients.

13.7 SUMMARY

As in Chapter 12 inferences from bivariate sample data are discussed, but in this chapter the case in which the variables are quantitative (rather than categorical) is covered.

The scatter diagram is a useful and important graphical summary of this type of data. A measure of the degree of association between the variables is provided by a correlation coefficient. If both variables are approximately normally distributed we may calculate:

$$(\text{Pearson's})\ r = \frac{\Sigma xy - \dfrac{\Sigma x \Sigma y}{n}}{\sqrt{\left\{\left[\Sigma x^2 - \dfrac{(\Sigma x)^2}{n}\right]\left[\Sigma y^2 - \dfrac{(\Sigma y)^2}{n}\right]\right\}}}.$$

In other cases we may calculate:

$$(\text{Spearman's})\ r_s = 1 - \frac{6\Sigma d^2}{n^3 - n}.$$

Hypothesis tests may be carried out for both Pearson's and Spearman's coefficients, the former being more powerful if both variables are normally distributed.

There are several important points to bear in mind when we try to interpret correlation coefficients.

WORKSHEET 13: CORRELATION COEFFICIENTS

Fill in the gaps in Questions 1–6.

1. If two quantitative variables are measured for a number of individuals the data may by plotted in a ________ ________.
2. A ________ ________ is a measure of the degree of association between two quantitative variables.
3. If it is reasonable to assume that each variable is normally distributed and we wish to obtain a measure of the degree of association between them, the appropriate ________ ________ to calculate is named ______'s and has the symbol ______. For the population the symbol is ______.
4. In calculating ______ we must decide which of our variables is the 'x' variable and which is the 'y' variable. The choice is ________.
5. The value of r (or r_s) must lie somewhere in the range ______ to ______. If the points on the scatter diagram indicate that as one variable increases the other variable tends to decrease the value of r will be ______. If the points show no tendency to either increase together or decrease together the value of r will be close to ________.
6. To decide whether there is a significant correlation between the two variables we carry out a hypothesis test for the population parameter ______ if the variables can be assumed to be approximately ______ ________. If we cannot make this assumption the null hypothesis is that the rankings of the two variables are ______.

7. The percentage increase in unemployment and the percentage increase in manufacturing output were recorded for a random sample of ten industrialized countries over a period of a year. The data are recorded below. Draw a scatter diagram. Is there a significant negative correlation? What further conclusions can be drawn, if any?

Percentage increase in unemployment	*Percentage increase in manufacturing output*
10	−5
5	−10
20	−12
15	−8
12	−4
2	−5
−5	−2
14	−15
1	6
−4	5

8. A company owns eight large luxury hotels, one in each of eight geographical areas. Each area has a different commercial television channel. To estimate the effect of television advertising the company carried out a month's trial in which the number of times a commercial, advertising the local luxury hotel, was shown was varied from one area to another. The percentage increase in the receipts of each hotel over the three months following the months's trial was also calculated:

Area	1	2	3	4	5	6	7	8
Number of times the commercial shown	0	0	0	10	20	30	40	50
Percentage increase in receipts	−2	5	10	5	7	14	13	11

What conclusions can be drawn?

9. In a mountainous region a drainage system consists of a number of basins with rivers flowing through them. For a random sample of seven basins, the area of each basin and the total length of the rivers flowing through each basin are as follows:

Basin number	*Area* (sq. km)	*River length* (km)
1	7	10
2	8	8
3	9	14
4	16	20
5	12	11
6	14	16
7	20	10

Are larger areas associated with longer river lengths?

10. From the data in the table that follows, showing the percentage of the population of a country using filtered water and the death rate due to typhoid for various years, calculate the correlation coefficient and test its significance at the 5% level. What conclusions would you draw about the cause of the reduction in the typhoid death rate from 1900 to 1912?

Year	*Percentage using filtered water*	*Typhoid death rate*/100 000 *living*
1900	9	36
1902	12	37
1904	16	35
1906	21	32
1908	23	27
1910	35	22
1912	45	14

11. A random sample of 20 families had the following annual income and annual savings in thousands of pounds:

Income	*Savings*
5.1	0.2
20.3	0.5
25.2	0.3
15.0	5.7
10.3	0.7
15.6	1.3

Income	*Savings*
16.0	0.4
7.3	4.2
8.6	2.0
12.3	0.6
14.0	0.3
8.9	0.1
12.4	0.0
14.0	0.7
16.0	0.2
14.0	0.3
15.3	1.0
12.4	0.6
10.3	0.5
11.3	0.7

Is there a significant positive correlation between income and savings?

Regression analysis

14.1 INTRODUCTION

When two quantitative variables are measured for a number of individuals we may be not so much interested in a measure of association between the variables (provided by a correlation coefficient) as in predicting the value of one variable from the value of the other variable. For example, if trainee salesmen take a test at the end of their training period, can the test score be used to predict the first-year sales, and how accurate is the prediction? One way to answer such a question is to collect both the test score and the first-year sales of a number of salesmen and from these sample data develop an equation relating these two variables. This equation is an example of a *regression equation*,[†] the simplest type of which is a *simple*[‡] *linear regression equation* which can be represented by a straight line on the scatter diagram for the two variables. We should be careful to draw the scatter diagram first, however, to decide whether the relationship between the variables appears to be reasonably linear (this is not the case in Fig. 13.3, for example).

The linear regression equation will be of the form

$$(\text{first-year sales}) = a + b \times (\text{test score})$$

where a and b are values we can calculate from the sample data.

In general, if we call the variable we wish to predict the y variable and the variable we use to do the predicting the x variable, the linear regression equation for y on x is

$$y = a + bx.$$

[†] The word 'regression' is one for which the original meaning is no longer useful. In the 19th century, Galton collected the heights of fathers and their sons and put forward the idea that, since very tall fathers tended to have slightly shorter sons, and very short fathers tended to have slightly taller sons, there would be what Galton termed a 'regression to the mean'.
[‡] The word 'simple' here implies that we are using only one variable (rather than many as in multiple regression) to predict another variable. Since this is the only case we will consider, the word simple will not be used again to describe regression equations.

The symbol b represents the slope (or gradient) of the line, and is also sometimes called the regression coefficient, and the symbol a represents the intercept.

14.2 DETERMINING THE REGRESSION EQUATION FROM SAMPLE DATA

Example

Suppose that for a random sample of eight salesmen their first-year sales and test scores as trainees are as in Table 14.1.

Table 14.1 First-year sales and test scores of eight salesmen

First-year sales (£ × 1000) y	*Test score* x
105	45
120	75
160	85
155	65
70	50
150	70
185	80
130	55

Notice we have labelled sales as the y variable we wish to predict from the x variable, namely test score.

The scatter diagram, Fig. 14.1, shows that there appears to be a linear relationship between the variables. We will now calculate a and b (in reverse order) using the formulae

$$b = \frac{\Sigma xy - \dfrac{\Sigma x \Sigma y}{n}}{\Sigma x^2 - \dfrac{(\Sigma x)^2}{n}} \quad \text{and} \quad a = \bar{y} - b\bar{x},$$

where $\bar{x}$ and $\bar{y}$ are the sample means of x and y, so

$$\bar{x} = \frac{\Sigma x}{n}, \qquad \bar{y} = \frac{\Sigma y}{n},$$

and n is the number of 'individuals'. For the data in Table 14.1.

$$\begin{aligned} \Sigma x &= 525 & \Sigma x^2 &= 35\,925 \\ \Sigma y &= 1\,075 & \Sigma y^2 &= 153\,575 \\ \Sigma xy &= 73\,350 & n &= 8. \end{aligned}$$

(Refer to similar calculations in Section 13.2 if necessary.)

$$b = \frac{73\,350 - \dfrac{525 \times 1075}{8}}{35\,925 - \dfrac{525^2}{8}} = \frac{2803}{1472} = 1.904$$

$$a = \frac{1075}{8} - 1.904 \times \frac{525}{8} = 134.4 - 125.0 = 9.4.$$

The regression equation of y (first-year score) on x (test score) is

$$y = 9.4 + 1.904x.$$

14.3 PLOTTING THE REGRESSION LINE ON THE SCATTER DIAGRAM

We can now plot the regression line (for the example in the previous section) to represent the regression equation, $y = 9.4 + 1.904x$. Since two points determine a straight line we can 'feed' any two values of x into the equation, calculate the corresponding values of y, and thus we have the co-ordinates of two points on the regression line. It is a good idea to use the minimum and maximum sample data values of x. (For the reason, see Note (b) below on extrapolation.)

The minimum value of x in Table 14.1 is 45. When $x = 45$, 'predicted y' $= 9.4 + 1.904 \times 45 = 95.1$. The maximum value of x is 85. When $x = 85$, 'predicted y' $= 9.4 + 1.904 \times 85 = 171.2$. We join the points

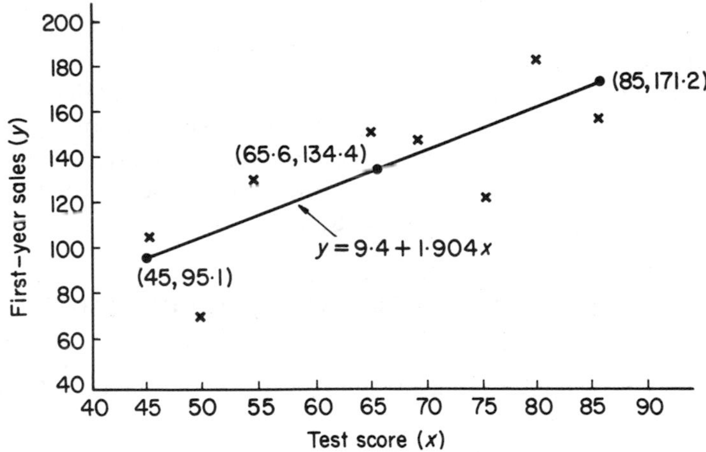

Fig. 14.1 Scatter diagram for first-year sales and test score.

(45, 95.1) and (85, 171.2) by a straight line on the scatter diagram, Fig. 14.1.

Notes

(a) The regression line is often referred to as the line of 'best fit'. In what sense is it best? The answer is that it is the line for which the sum of squares of the distances from the points to the line in the y direction is minimized. There are good theoretical reasons for using this criterion. The formulae for a and b which we used were derived by calculus using this criterion, but the derivation is beyond the scope of this book.

(b) The regression line and the regression equation apply only within the range of the x values in the data. The line should not be extrapolated (extended) much below the minimum value of x or much above the maximum value of x. Nor should we use the regression equation for values of x outside the range of the data.

(c) As a check on the position of the line on the scatter diagram, it can be shown that the line should pass through the point $(\bar{x}, \bar{y})$. For the example this point is (65.5, 134.4).

14.4 PREDICTING VALUES OF y

For a particular value of x (within the range of the sample data) the corresponding point on the line gives the predicted value of y. This may also be obtained, and with more accuracy, by feeding the particular value of x into the regression equation.

Example

Predict first-year sales for a test score of 60.

For $x = 60$, 'predicted y' $= 9.4 + 1.904 \times 60 = 123.6$. What does this mean? It is an estimate of the mean first-year sales of salesmen with test scores of 60.

Using the ideas of Chapter 9 we may also calculate a confidence interval for the mean sales which we would predict for test scores of 60 (and any other test scores within the range of the sample data) to give an idea of the precision of the estimate of first-year sales for test scores of 60. These calculations are shown in the next section.

*14.5 CONFIDENCE INTERVALS FOR PREDICTED VALUES OF y

The formula for a 95% confidence interval for the predicted value of y at some value $x = x_0$ is:

$$(a + bx_0) \pm ts_r \sqrt{\left[\frac{1}{n} + \frac{(x_0 - \bar{x})^2}{\Sigma x^2 - \frac{(\Sigma x)^2}{n}}\right]}$$

where

$$s_r^2 = \frac{\left[\Sigma y^2 - \frac{(\Sigma y)^2}{n}\right] - b^2\left[\Sigma x^2 - \frac{(\Sigma x)^2}{n}\right]}{n - 2}$$

and t is obtained from Table D.5 for $\alpha = 0.025$, $\nu = n - 2$.

Assumptions in using this formula The data points are distributed approximately normally about the regression line in the y direction. The distribution is the same for all values of x, as in Fig. 14.2.

The assumption of approximate normality is less important the larger the number of individuals, n. The assumption of a constant normal distribution mainly requires that the variability about the line in the y direction is the same all along the line, and does not, for example, tend to increase (or decrease) significantly as x increases.

Example

For the example of the salesmen, we can use the confidence interval formula for values of x_0 between 45 and 85.

$$s_r^2 = \frac{\left[153\,575 - \frac{1075^2}{8}\right] - 1.904^2\left[35\,925 - \frac{525^2}{8}\right]}{8 - 2}$$

$$= \frac{9122 - 1.904^2 \times 1472}{6}$$

$$= \frac{3786}{6}$$

$$= 631,$$

$$s_r = 25.1.$$

Notes

(a) The numerator of the formula for s_r^2 represents the sum of squares of the distances from the points to the line in the y direction. This sum of squares is 3786 for the data in the example. For any other line drawn on the scatter diagram the sum of squares will exceed 3786.

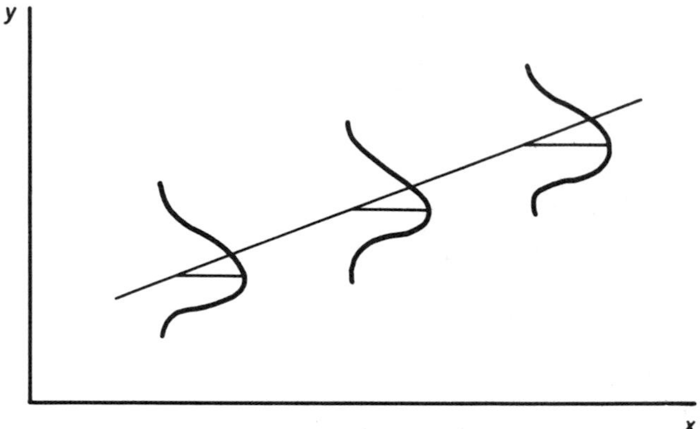

Fig. 14.2 Assumptions required in using formula for confidence intervals for predicted y.

(b) s_r is called the residual standard deviation, and the distances from the points to the line in the y direction are called residuals (see Fig. 14.3).

At $x_0 = 60$, the 95% confidence interval for predicted y is

$$(9.4 + 1.904 \times 60) \pm 2.447 \times 25.1 \sqrt{\left[\frac{1}{8} + \frac{(60 - 65.6)^2}{35\,925 - \dfrac{525^2}{8}}\right]}$$

$$123.6 \pm 23.5$$
$$100.1 \text{ to } 147.1.$$

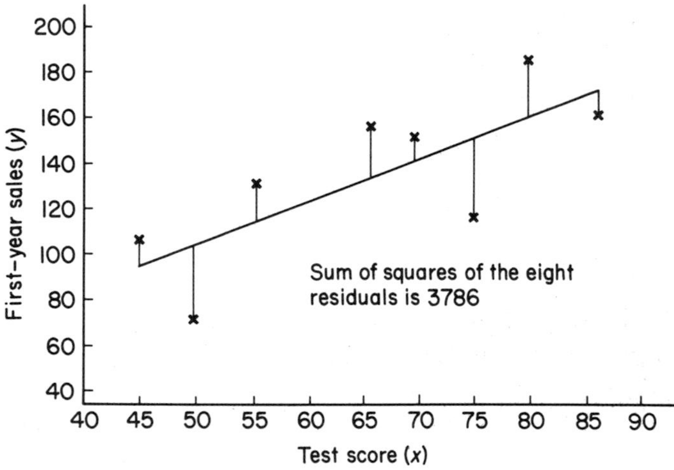

Fig. 14.3 The residuals for the data from Table 14.1.

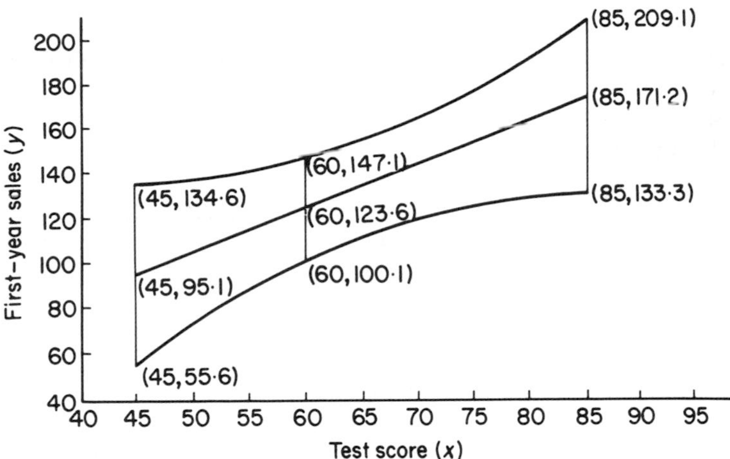

Fig. 14.4 95% confidence intervals for predicting y, data from Table 14.1.

For a test score of 60, then, the mean first-year sales will be between 100.1 and 147.1, for 95% confidence.

For other test scores, similar calculations can be made, and are represented in Fig. 14.4.

The 95% confidence interval is narrowest at $x = 65.6$, the mean value of x, and widens as we move away from the mean in either direction (increasing or decreasing x). This result agrees with the intuitive idea that we expect less error in our prediction the nearer we are to the middle of the data, and conversely we expect a larger error when we move to the extremes of the data.

*14.6 HYPOTHESIS TEST FOR THE SLOPE OF THE REGRESSION LINE

For the example we calculated $b = 1.904$ (in Section 14.2). This is the slope of the regression line for the sample data. It is our estimate of the increase in y (first-year sales) for unit increase in x (test score).

We can also conceive of a regression line for a population of salesmen with an equation:

$$y = \alpha + \beta x$$

where β is the slope, α is the intercept. So our estimate of β is the sample estimate provided by b, namely 1.904. Could such a sample value have arisen if the population value of β had been zero (implying a horizontal population regression line)? To answer this we carry out a hypothesis test as follows:

1. $H_0 : \beta = 0$.
2. $H_1 : \beta > 0$, one-sided, implying that higher test scores result in higher first-year sales.
3. 5% significance level.
4. The formula to use for this test is

$$Calc\ t = \frac{b}{s_r \Big/ \sqrt{\left[\Sigma x^2 - \frac{(\Sigma x)^2}{n}\right]}} = \frac{1.904}{25.1 \Big/ \sqrt{\left[35\,925 - \frac{525^2}{8}\right]}}$$

for the example

$$= 2.91.$$

5. $Tab\ t = 1.943$, for $\alpha = 0.05/1 = 0.05$, $v = (n - 2) = 6$.
6. Since $Calc\ t > Tab\ t$, reject H_0.
7. The slope of the regression line is significantly greater than zero (5% level).

   Assumptions The same assumptions apply as in Section 14.5.

*14.7 THE CONNECTION BETWEEN REGRESSION AND CORRELATION

In the previous section, had we not rejected the null hypothesis $H_0 : \beta = 0$ then it would have been reasonable to regard the population regression line as being horizontal, in other words parallel to the x axis. So that, whatever the value of x, the predicted value of y would have been the

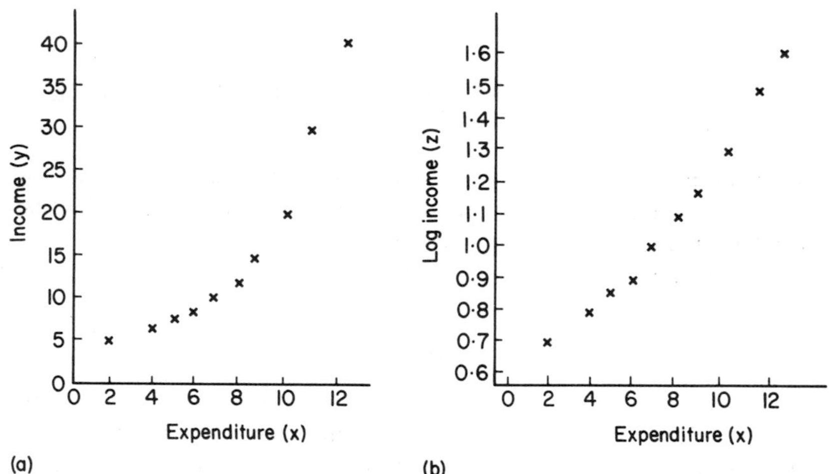

Fig. 14.5 Scatter diagrams for (a) income v. expenditure on meat; (b) log income v. expenditure on meat.

same. Another way of expressing this idea is to describe the variables as uncorrelated.

If $H_0 : \beta = 0$ is rejected in favour of $H_1 : \beta > 0$ then higher values of x would give rise to higher predicted values of y, and we would say the variables were positively correlated.

This intuitive reasoning is supported by the fact that an hypothesis test for Pearson's ϱ (see Section 13.3) would result in values of *Calc t* and *Tab t* identical to those found in Section 14.6. This statement is left for you to confirm by calculation.

*14.8 TRANSFORMATIONS TO PRODUCE LINEAR RELATIONSHIPS

When the scatter diagram shows a non-linear pattern it may be possible to produce a linear pattern by 'transforming' one of the variables.

Example

The annual income (in thousands of pounds) and the annual expenditure on meat (hundreds of pounds) by a random sample of ten families of the same size and average age are shown in Table 14.2.

The x and y variables are as in the table, and a new variable z has been calculated by taking log to base 10 of the y values. The new variable z is an example of a transformation of the variable y. The reason for introducing the new variable is clear if we study the two scatter diagrams of Fig. 14.5. The plot of y against x shows a tendency to be non-linear, but the plot of z against x show a stronger linear relationship between the latter pair of variables. If we wish to predict income from expenditure on meat this may be achieved by calculating the regression equation for z on x, and then using the fact that $z = \log_{10} y$.

For $z = a + bx$, and using the summations

$$\Sigma x = 74, \quad \Sigma x^2 = 640, \quad \Sigma z = 10.87, \quad \Sigma xz = 88.91 \quad n = 10,$$

$b = 0.0917$ and $a = 0.4085$ are obtained. So $z = 0.4085 + 0.0917x$, which we can also write as:

$$\log_{10} y = 0.4085 + 0.0917x.$$

This equation may be used to predict y from x. The equation may be represented as a straight line on the plot of z against x, but will be a curve on the plot of y against x.

The choice of what transformation to use in a particular case is beyond the scope of this book. Apart from the 'log transformation', other transformations (which may be appropriate depending on the type of data collected) are square roots (for Poisson data) or arcsines (for binomial data).

Table 14.2 Income and expenditure on meat for 10 families

Annual income $y(£ \times 1000)$	*Annual expenditure on meat* $x(£ \times 100)$	$z = \log_{10} y$
5	2	0.70
6	4	0.78
7	5	0.85
8	6	0.90
10	7	1.00
12	8	1.08
15	9	1.18
20	10	1.30
30	11	1.48
40	12	1.60

14.9 SUMMARY

(Simple) linear regression analysis is a method of deriving an equation relating two quantitative variables. The equation, which may be represented by a straight line on a scatter diagram, is:

$$y = a + bx, \quad \text{where } b = \frac{\Sigma xy - \dfrac{\Sigma x \Sigma y}{n}}{\Sigma x^2 - \dfrac{(\Sigma x)^2}{n}} \quad \text{and} \quad a = \bar{y} - b\bar{x}.$$

The equation is used to predict values of y for values of x within the range of sample values.

Under certain assumptions, confidence intervals for predicted values of y may be calculated, the null hypothesis that the slope of the population regression line, β, is zero may also be tested, and a confidence interval for β may be calculated (see Worksheet 14, Question 8).

The connection between regression and correlation was discussed, as was an example of transforming one of the variables to 'linearize' a non-linear pattern in the scatter diagram.

WORKSHEET 14: REGRESSION ANALYSIS

Fill in the gaps in Questions 1 and 2.

1. The purpose of (simple) regression analysis is to ______ values of one variable for particular values of another variable. We call the variable

whose values we wish to predict the ______ variable, and the other we call the ______ variable.

2. Using sample values of the two variables the ______ diagram is drawn. If this appears to show a linear relationship between the variables we calculate a and b for the linear ______ equation. This equation may be represented as a ______ ______ on the scatter diagram.
3. The regression line is also called the line of 'best-fit' because it minimizes the sum of squares of the distances from the points to the line in the y direction. For the example in Section 14.2, this sum of squares is 3786. Draw any other line 'by eye' on the scatter diagram which you think might be a better fit and calculate the sum of squares for your line. You should not be able to beat 3786, rounding errors excepted.
4. The following table gives the number of bathers at an open-air swimming pool and the maximum recorded temperature on ten Saturdays during one summer:

Number of bathers	*Maximum temperature* (°C)
290	19
340	23
360	20
410	24
350	21
420	26
330	20
450	25
350	22
400	29

Draw a scatter diagram and calculate the slope and intercept of the regression line which could be used to predict the number of bathers from the maximum temperature. Plot the regression line on the scatter diagram, checking that your line passes through the point $(\bar{x}, \bar{y})$.

How many bathers would you predict if the forecast for the maximum temperature on the next Saturday in the summer was:

(a) 20°C, (b) 25°C, (c) 30°C?

Which of these predictions will be the least reliable?

5. In order to estimate the depth of water beneath the keel of a boat a sonar measuring device was fitted. The device was tested by observing the sonar readings over a number of known depths, and the following data were collected:

Sonar reading (metres)	0.15	0.91	1.85	3.14	4.05	4.95
True depth of water (metres)	0.2	1	2	3	4	5

Draw a scatter diagram for these data and derive a linear regression equation which could be used to predict the true depth of water from the sonar reading. Predict the true depth for a sonar reading of zero and obtain a 95% confidence interval for your prediction. Interpret your result.

6. The moisture content of a raw material and the relative humidity of the atmosphere in the store where the raw material is kept were measured on seven randomly selected days. On each day one randomly selected sample of raw material was used.

Relative humidity (%)	30	35	52	38	40	34	60
Moisture content (%)	7	10	14	9	11	6	16

Draw a scatter diagram and derive a linear regression equation which could be used to predict the moisture content of the raw material from the relative humidity. Use the equation to predict moisture content for a relative humidity of:

(a) 0%, (b) 50%, (c) 100%.

Also test the hypothesis that the slope of the (population) regression line is zero.

7. The data below give the weight and daily food consumption for 12 obese adolescent girls. Calculate the best-fit linear regression equation which would enable you to predict food consumption from weight, checking initially that the relationship between the variables appears to be linear.

Weight (kg)	85	95	80	60	95	85	90	80	85	70	65	75
Food consumption (100s of calories/day)	32	33	33	24	39	32	34	28	33	27	26	29

What food consumption would you predict, with 95% confidence, for adolescent girls weighing: (a) 65, (b) 80, (c) 95 kg?

8. To see if there is a linear relationship between the size of boulders in a stream and the distance from the source of the stream, samples of boulders were measured at 1 kilometre intervals. The average sizes of boulders found at various distances were as follows:

Distance downstream (km)	1	2	3	4	5	6	7	8	9	10
Average boulder size (cm)	105	85	80	85	75	70	75	60	50	55

Find the regression equation which could be used to predict average boulder size from distance downstream. Plot the regression line on the scatter diagram. Test the null hypothesis $\beta = 0$ against the alternative hypothesis $\beta < 0$. Also obtain a 95% confidence interval for β using the formula:

$$b \pm \frac{ts_r}{\sqrt{\left[\Sigma x^2 - \frac{(\Sigma x)^2}{n}\right]}}$$

where t is from Table D.5 for $\alpha = 0.025$, and $v = n - 2$.

9. The number of grams of a given salt which will dissolve in 100 g of water at different temperatures is shown in the table below:

Temperature (°C)	0	10	20	30	40	50	60	70
Weight of salt (g)	53.5	59.5	65.2	70.6	75.5	80.2	85.5	90.0

Find the regression equation which can be used to predict weight of salt from temperature. Plot the regression line on the scatter diagram. Predict the weight of salt which you estimate would dissolve at temperatures of:

(a) 25°C (b) 55°C (c) 85°C

and comment on your results.

10. A random sample of ten people who regularly attempted the daily crossword puzzle in a particular national newspaper were asked to

time themselves on a puzzle which none of them had seen before. Their times to complete the puzzle and their scores in a standard IQ test were as follows:

IQ	120	100	130	110	100	140	130	110	150	90
Times (minutes)	9	7	13	8	4	5	16	7	5	13

What conclusions can you draw from these data?

*11. The following data show the values of two variables x and y obtained in a laboratory experiment on seven rats:

x	0.4	0.5	0.7	0.9	1.3	2.0	2.5
y	2.7	4.4	5.4	6.9	8.1	8.4	8.6

It is thought that there is a linear relationship between either (a) y and x or (b) y and $1/x$. Plot y against x, and y against $1/x$ and decide from the two scatter diagrams which will result in a better linear relationship. Calculate the coefficients of the appropriate linear regression equation and plot the regression line on the scatter diagram. Obtain the predicted value of y for $x = 1$ and calculate a 95% confidence interval for your prediction.

Without performing any further calculations, decide which of the following will be the greater:

(i) The Pearson correlation coefficient between y and x.

(ii) The Pearson correlation coefficient between y and $1/x$.

χ^2 goodness-of-fit tests

15.1 INTRODUCTION

We now return to a one-variable problem, namely the problem of deciding whether the sample data for one variable could have been selected from a particular type of distribution. Four types will be considered:

Type of distribution	*Type of variable*
'Simple proportion'	Categorical
Binomial	Discrete
Poisson	Discrete
Normal	Continuous

In each case a χ^2 test will be used to see how closely the frequencies of the observed sample values agree with the frequencies we would expect under the null hypothesis that the sample data actually do come from the type of distribution being considered (refer to Chapter 12 now if you are unfamiliar with the χ^2 test).

15.2 GOODNESS-OF-FIT FOR A 'SIMPLE PROPORTION' DISTRIBUTION

We define a simple proportion distribution as one for which we expect the frequencies of the various categories, into which the 'values' of a categorical variable will fall, to be in certain numerical proportions.

Example

The ratio of numbered cards to picture cards in a pack is 36 to 16, which we could write as 36 : 16. If we selected cards randomly with replacement we would expect the proportions of numbered and picture cards to be 36/52 and 16/52 respectively.

Example

Suppose that there is a genetic theory that adults should have hair colours of black, brown, fair and red in the ratios 5 : 3 : 1 : 1. If this theory is correct we expect the frequencies of black, brown, fair and red hair to be in the proportions:

$$\frac{5}{5+3+1+1}, \quad \frac{3}{5+3+1+1}, \quad \frac{1}{5+3+1+1}, \quad \frac{1}{5+3+1+1}$$

or 5/10, 3/10, 1/10, 1/10.

If we take a random sample of 50 people to test this theory, we would expect 25, 15, 5, 5 to have black, brown, fair and red hair respectively (we simply multiply the expected proportions by the sample size). We can then compare these expected frequencies with the frequencies we actually observed in the sample and calculate a χ^2 statistic.

It is convenient to set this calculation in the form of a table (see Table 15.1).

Table 15.1 Calculation of χ^2 for a 5 : 3 : 1 : 1 distribution

Hair colour	*Expected proportions*	*Expected frequencies (E)*	*Observed frequencies (O)*	$\frac{(O-E)^2}{E}$
Black	$\frac{5}{10}$	25	28	0.36
Brown	$\frac{3}{10}$	15	12	0.60
Fair	$\frac{1}{10}$	5	6	0.20
Red	$\frac{1}{10}$	5	4	0.20
		50	50	*Calc* $\chi^2 = 1.36$

Notice that the method of calculating the E values ensures that the sum of the E values equals the sum of the O values.

We now set out the seven-step hypothesis test for this example:

1. H_0: Genetic theory is correct. Sample data do come from a 5 : 3 : 1 : 1 distribution.
2. H_1: Genetic theory is not correct (two-sided).
3. 5% significance level.
4. $$Calc\ \chi^2 = \Sigma\frac{(O-E)^2}{E} = 1.36,$$

   from Table 15.1.

5. *Tab* $\chi^2 = 7.82$ for α = sig level = 0.05 (even though H_1 is two-sided) and ν = (number of categories − 1) = 4 − 1 = 3, from Table D.8.
6. Since *Calc* $\chi^2 <$ Tab χ^2, do not reject H_0.
7. It is reasonable to conclude that the genetic theory is correct (5% level).

Notes

(a) The formula for *Calc* χ^2 is only valid if all the E values are $\geqslant 5$. If any E value is <5, it may be sensible to combine categories so that all E values for the new categories are $\geqslant 5$.

(b) The formula

$$Calc\ \chi^2 = \Sigma \frac{(O - E)^2}{E}$$

is used if $\nu > 1$. If $\nu = 1$ use

$$Calc\ \chi^2 = \Sigma \frac{(|O - E| - \frac{1}{2})^2}{E}$$

(applying Yates's correction as in Section 12.3).

(c) The formula for degrees of freedom, ν = (number of categories − 1), may be justified by reference to Section 9.7, the one restriction being that the sum of the E values must equal the sum of the O values. Only three of the E values may be determined independently in the example, so there are three degrees of freedom.

(d) If some categories are combined (see (a) above), the number of categories to be used to calculate the degrees of freedom is the number after combinations have been made.

*15.3 GOODNESS-OF-FIT FOR A BINOMIAL DISTRIBUTION

Suppose we carry out n trials where each trial can result in one of only two possible outcomes, which we call success and failure. Suppose we repeat this set of n trials several times and observe the frequencies for the various numbers of successes which occur. We may then carry out a χ^2 test to decide whether it is reasonable to assume that the number of successes in n trials has a binomial distribution with a 'p value' which we estimate from the observed frequencies, or we may carry out a similar test in which we specify the particular value of p without reference to the observed frequencies. (It will be assumed that you are familiar with the binomial distribution as described in Chapter 6.)

Example

In an experiment in extra-sensory perception (ESP) four cards marked A, B, C and D are used. The experimenter, unseen by the subject, shuffles the cards and selects one. The subject tries to decide which card has been selected. This procedure is repeated five times for each of a random sample of 50 subjects. The number of times out of five when each subject correctly identifies a selected card is counted. Suppose the data for all 50 subjects are recorded in a frequency distribution (see Table 15.2).

Table 15.2 Results of an ESP experiment, 50 subjects, 5 trials per subject

Number of correct decisions	0	1	2	3	4	5
Number of subjects	15	18	8	5	3	1

Is there evidence that subjects are simply guessing? We may regard the testing of each subject as a set of five trials, each trial having one of two possible outcomes, 'correct decision' or 'incorrect decision'. This set of five trials is repeated (on different subjects) a total of 50 times. The second row in Table 15.2 gives the observed frequencies (O) for the various possible numbers of correct decisions.

If subjects are guessing then the probability of a correct decision is 0.25 for each selection, since the four cards are equally likely to be selected. The question above, 'Is there evidence that subjects are simply guessing?' is equivalent to the question:

> Is it reasonable to suppose that the data in Table 15.2 come from a binomial distribution with $n = 5$, $p = 0.25$?

The expected frequencies (E) for the various numbers of correct decisions are obtained by assuming, for the purposes of the calculation, that we are dealing with a binomial distribution with $n = 5, p = 0.25$. First we calculate the probabilities of 0, 1, 2, 3, 4 and 5 correct decisions (using the methods of Chapter 6). These probabilities are multiplied by the total of the observed frequencies (50 in the example) to give the expected frequencies.

These calculations and the calculation of χ^2 are set out in Table 15.3.

Table 15 3 Calculation of χ^2 for a binomial distribution

Number of correct decisions (x)	P(x)	$E = \text{P}(x) \times 50$		O			$\frac{(O - E)^2}{E}$
0	0.2373	11.9		15			0.81
1	0.3955	19.8		18			0.16
2	0.2637	13.2		8			2.05
3	0.0879	4.4	} 5.2	5	} 9		2.78
4	0.0146	0.7		3			
5	0.0010	0.1		1			
Total	1.0000	50.1		50		*Calc* χ^2 =	5.80

Notes

(a) The probabilities P(x) were obtained from Table D.1 for n = 5, p = 0.25.

(b) Because three E values are <5, the bottom three categories have been combined.

(c) The totals of the E and O columns are equal (apart from rounding errors).

We now set out the seven-step hypothesis test for this example:

1. H_0: Sample data do come from a $B(5, 0.25)$ distribution, implying that the subjects are guessing.
2. H_1: Sample data do not come from a $B(5, 0.25)$ distribution, which might imply that some subjects have powers of ESP.
3. 5% significance level.
4. *Calc* χ^2 = 5.80, from Table 15.3.
5. *Tab* χ^2 = 7.82, for α = 0.05, ν = number of categories − 1 = 3. (Refer to Notes (a) and (b) below, and Table D.8).
6. Since Calc χ^2 < *Tab* χ^2, do not reject H_0.
7. It is reasonable to suppose subjects are guessing (5% level).

Notes

(a) Although there were six categories initially, there are only four after combinations.

(b) In cases where the value of p is not specified by the experimental set-up, unlike the example above, we have to estimate it from the data of observed frequencies (see Worksheet 15, Question 5) and we lose a further degree of freedom.

*15.4 GOODNESS-OF-FIT FOR A POISSON DISTRIBUTION

Suppose we observe the number of times a particular event occurs in each of a number of units of time (or space). Can we conclude that the number of occurrences of the event per unit time (or space) has a Poisson distribution, implying randomly occurring events?

Example

Suppose that the number of major earthquakes occurring per month is collected for 100 months in a frequency distribution (see Table 15.4).

Table 15.4 Number of earthquakes occurring in 100 months

Number of earthquakes per month	0	1	2	3	4
Number of months	57	31	8	3	1

The observed frequencies (O) for the various numbers of earthquakes per month are given in the second row of Table 15.4.

The expected frequencies (E) for the various numbers of earthquakes are obtained by assuming, for the purposes of the calculation, that we are dealing with a Poisson distribution. The parameter m, the mean of the distribution, is estimated by the sample mean number of earthquakes per month:

$$= \frac{\text{total number of earthquakes}}{\text{total number of months}}$$

$$= \frac{57 \times 0 + 31 \times 1 + 8 \times 2 + 3 \times 3 + 1 \times 4}{100}$$

$$= 0.6.$$

For the Poisson distribution with a mean $m = 0.6$ we calculate the probabilities of 0, 1, 2, 3 and 4 or more earthquakes (using the methods of Chapter 6). These probabilities are multiplied by the total of the observed frequencies (100 in the example) to give the expected frequencies.

These calculations and the calculation of χ^2 are set out in Table 15.5.

Table 15.5 Calculation of χ^2 for a Poisson distribution

Number of earthquakes per month (x)	P(x)	E = P(x) × 100	O	$\frac{(O-E)^2}{E}$
0	0.5488	54.9	57	0.08
1	0.3293	32.9	31	0.11
2	0.0988	9.9 }	8 }	
3	0.0197	2.0 } 12.2	3 } 12	0.00
4 or more	0.0034	0.3 }	1 }	
Total	1.0000	100.0	100	*Calc* χ^2 = 0.19

Notes

(a) The probabilities P(x) were obtained from Table D.2 for $m = 0.6$ (see Section 6.12).

(b) The probability of 4 or more (rather than 4) is calculated to ensure that the totals of the E and O columns are equal (apart from rounding errors).

(c) The bottom three categories have been combined because two E values are <5.

We now set out the seven-step hypothesis test for this example:

1. H_0: Sample data come from a Poisson distribution, implying earthquakes occur randomly in time.

2. H_1 : Sample data do not come from a Poisson distribution.
3. 5% significance level.
4. *Calc* χ^2 = 0.19, from Table 15.5.
5. *Tab* χ^2 − 3.84 for $\alpha = 0.05$, ν = number of categories − 1 − 1 = 1 (see note below).
6. Since *Calc* χ^2 < *Tab* χ^2, do not reject H_0.
7. It is reasonable to assume a Poisson distribution, and that earthquakes occur randomly in time (5% level).

Note

There are only three categories after combinations. One degree of freedom is lost because of the restriction, $\Sigma E = \Sigma O$, and another is lost because the parameter m is estimated from the sample data.

*15.5 GOODNESS-OF-FIT FOR A NORMAL DISTRIBUTION

Many of the analyses carried out in earlier chapters required the assumption of at least approximate normality for the continuous variable of interest. When we have a reasonably large sample size[†] a histogram may give some indication of whether the distribution is normal, but a χ^2 test provides a more objective way of testing for normality in this case.

Example

Three hundred staff in an office block were asked how far they lived from work. The data are summarized in a group frequency distribution (see Table 15.6).

Table 15.6 Distance from home to work for 300 staff

Distance from home to work (miles)	*Number of staff*
Less than 5	60
Between 5 and 10	100
Between 10 and 15	75
Between 15 and 20	45
Between 20 and 25	15
Between 25 and 30	4
Between 30 and 35	1

[†] *Note* For a test of normality for small samples size, the Shapiro–Wilk W test may be used. Refer to Wetherill (1981) listed in Appendix E.

A histogram would show some positive skewness, but is this sufficient to conclude that the distribution is non-normal?

The observed frequencies (O) of the 'distance from home to work' are given in the second column of Table 15.6. The expected frequencies (E) of the distance from home to work are obtained by assuming, for the purposes of the calculation, that we are dealing with a normal distribution. The parameters μ and σ (the mean and standard deviation) for this distribution are estimated by the sample mean and sample standard deviation ($\bar{x}$ and s) for the sample data in Table 15.6.

Using the methods of Sections 4.2 and 4.7, we obtain, $\bar{x} = 10.4$ miles, $s = 6.1$ miles. For the normal distribution with a mean $\mu = 10.4$, standard deviation $\sigma = 6.1$ we calculate the probabilities for the various groups in Table 15.6 (refer to Section 7.3 if necessary). These probabilities may also be represented as areas under a normal distribution curve (see Fig. 15.1).

The probabilities are multiplied by the total of the observed frequencies (300 in the example) to give the expected frequencies. The calculations and the calculation of χ^2 are set out in Table 15.7.

Notes

(a) In the column labelled 'Distance', the ranges correspond to those in Table 15.6, except for 'greater than 30' which replaces '30 – 35' to ensure that the totals of the E and O columns are equal (except for rounding errors).

(b) The values in the column labelled 'Probability' are obtained using the methods of Section 7.3

(c) The bottom three categories have been combined because two E values are <5.

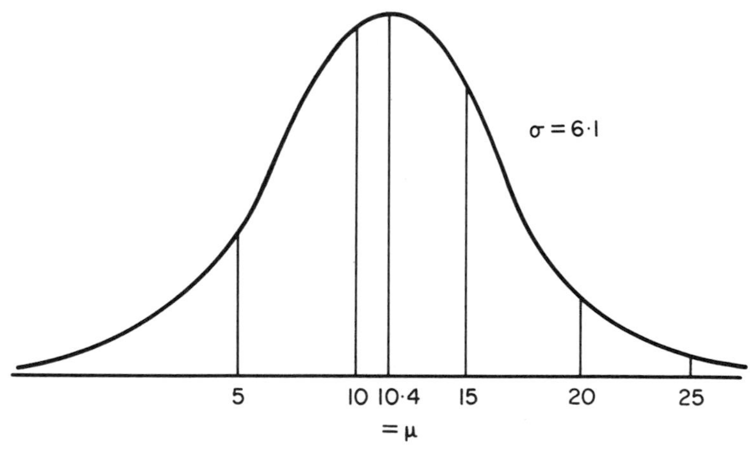

Fig. 15.1 A normal distribution data $\mu = 10.4$, $\sigma = 6.1$.

Table 15.7 Calculation of χ^2 for a normal distribution

Distance from home to work x	$z = \frac{x - 10.4}{6.1}$	*Areas from Table D.3(a)*	*Distance*	*Probability*	E (= Prob. × 300)	O	$\frac{(O-E)^2}{E}$
			<5	0.1867	56.0	60	
5	−0.89	0.8133					
			5 – 10	0.2854	85.6	100	0.29
10	−0.07	0.5279					
			10 – 15	0.3013	90.4	75	2.42
15	+0.75	0.7734					
			15 – 20	0.1684	50.5	45	2.62
20	+1.57	0.9418					
			20 – 25	0.0498	14.9	15	0.60
25	+2.39	0.9916					
			25 – 30	0.0077	2.3 (17.4)	4 (20)	
30	+3.21	0.9993					
			>30	0.0007	0.2	1	0.39
			Total	1.0000	299.9	300	*Calc* $\chi^2 = 6.32$

We now set out the seven-step hypothesis test for this example:

1. H_0 : Sample data come from a normal distribution.
2. H_1 : Sample data do not come from a normal distribution.
3. 5% significance level.
4. *Calc* $\chi^2 = 6.32$, from Table 15.7.
5. *Tab* $\chi^2 = 5.99$, for $\alpha = 0.05$, $\nu = 5 - 1 - 2 = 2$ (see note below).
6. Since *Calc* $\chi^2 >$ *Tab* χ^2, reject H_0.
7. It is not reasonable to assume that the distribution is normal (5% level).

Note

There are five categories after combinations. One degree of freedom is lost because of the restriction, $\Sigma E = \Sigma O$, and two more are lost because the two parameters μ and σ are estimated from the sample data.

15.6 SUMMARY

Goodness-of-fit tests using the χ^2 statistic are tests to decide whether it is reasonable to conclude that a sample of univariate (one-variable) data could have been drawn from a particular type of distribution. Four types of distribution are covered:

Simple proportion
Binomial
Poisson
Normal.

The observed data are in the form of frequency distributions and the expected frequencies are calculated assuming, for the purpose of the calculation, that the sample data come from the particular type of distribution being investigated. The degrees of freedom for *Tab* χ^2 are, in general, equal to (number of categories after combinations) − (number of parameters of the distribution estimated from the sample data) − 1.

WORKSHEET 15: χ^2 GOODNESS-OF-FIT TESTS

1. In Mendel's experiments with peas he classified 556 peas into four categories as follows:

Type of pea	*Number of peas*
Round and yellow	315
Round and green	108
Wrinkled and yellow	101
Wrinkled and green	32

Are these data consistent with his theory of heredity that these categories should occur in the proportions 9 : 3 : 3 : 1?

2. The number of fatal road accidents in one year in a large city were tabulated according to the time of occurrence:

Time	00.00–04.00	04.00–08.00	08.00–12.00	12.00–16.00	16.00–20.00	20.00–24.00
Number of accidents	28	15	14	18	15	30

Test the hypotheses that:
(a) Accidents are uniformly distributed in time.
(b) Accidents occur in the ratios 2 : 1 : 5 : 4 : 5 : 3, these being the estimated ratios of the volumes of traffic occurring in the city for the six four-hour periods.

3. The number of sheep farms of a given size in a county and the type of land on which they were situated were as follows:

Type of land	*Number of sheep farms*
Flat	43
Hilly	32
Mountainous	5

If 35% of the county is flat, 50% is hilly and 15% is mountainous, is the number of farms independent of the type of land?

4. A random sample of 100 families were asked how many cars they owned. The results were:

Number of cars	0	1	2 or more
Number of families	35	45	20

Test the hypothesis that, for all families, the ratios are 1 : 2 : 1 for the three categories of the number of cars owned.

*5. For a random sample of 300 families each with three children, the distribution of the number of boys was as follows:

Number of boys	0	1	2	3
Number of families	55	108	102	35

Test the hypotheses that the number of boys in families with three children has:

(a) A binomial distribution with a p value of 0.5.
(b) A binomial distribution. *Hint*: estimate p from sample data using the relative frequency definition, that is:

$$\frac{\text{total number of boys in the 300 families}}{\text{total number of children in the 300 families}}.$$

Compare the conclusions of (a) and (b).

*6. Samples of 10 pebbles were taken at random from each of 200 randomly selected sites on a beach. The number of limestone pebbles in each sample was counted. The results were summarized in a frequency distribution table:

Number of limestone pebbles	0	1	2	3	4	5	6	7	8	9	10
Number of sites	0	7	20	45	53	39	25	8	3	0	0

How would you have selected the sites randomly? How would you select a sample of 10 pebbles from a particular site? Is it reasonable to conclude that the number of limestone pebbles in samples of 10 has a binomial distribution with a p value of 0.4?

*7. An experiment was carried out to test whether the digit '8' occurred randomly in tables of random numbers. Successive sets of 20 single digits (0, 1, 2,. . ., 9) were examined and the number of times the digit '8' occurred was noted for each set.

Number of '8' digits found	0	1	2	3	4	5	6 or more
Number of sets	25	45	70	35	15	10	0

What conclusions can be drawn from these data?

*8. A survey was conducted to decide whether a particular plant species was randomly distributed in a meadow. Eighty points in the meadow were randomly selected. A quadrat was placed with its centre at each selected point and the number of individual plants of the species was noted:

Number of individual plants per quadrat	0	1	2	3	4	5	6
Number of quadrats	11	37	12	7	6	4	3

How would you have selected 80 points randomly in a meadow? Is it reasonable to assume that the plant species was randomly distributed in the meadow?

*9. The number of dust particles occurring in unit volume of a gas was counted. The procedure was repeated 100 times for the same constant volume. Given the following results, is it reasonable to assume that the number of dust particles per unit volume is randomly distributed with a mean of two particles per unit volume?

Number of particles	0	1	2	3	4 or more
Number of times this number of particles observed	9	32	26	15	18

*10. The number of minor defects noted by an inspector in 90 cars leaving a production assembly line was as follows:

Number of defects	0	1	2	3	4	5	6
Number of cars	35	13	6	5	18	10	3

(a) Test first whether the mean and variance of the number of defects are approximately equal. (This is a quick but not very reliable test for a Poisson distribution.)
(b) Now use the χ^2 test to decide whether the number of defects per car is randomly distributed.

*11. A nurseryman took a random sample of 50 plants from one of his beds a few weeks after they had germinated and measured their heights. Test whether a normal distribution could reasonably be a 'model' for these data:

Height (cm)	5–10	10–15	15–20	20–25	25–30	30–35
Number of plants	3	6	14	18	8	1

*12. Decide whether the weights of coffee in jars marked 200 g are normally distributed using the data from Worksheet 3, Question 2(a).

*13. The scores of 150 students who took a verbal scholastic aptitude test were as follows:

Score	*Number of students*
0–9	2
10–19	13
20–29	9
30–39	10
40–49	12
50–59	18
60–69	40
70–79	27
80–89	16
90–99	3
100	0

Describe the main features of this distribution, and test for normality. (Treat score as a continuous variable, so that the group 10–19 is 9.5–19.5 on a continuous scale with a mid-point of 14.5, and so on.)

16

Minitab

16.1 INTRODUCTION

Minitab is an interactive statistical computer package which is widely employed in colleges as an aid to teaching statistics. The purpose of this chapter is to give examples of the use of Minitab which relate to the statistical methods discussed so far in this book.

No attempt has been made to give a detailed account of the package. For such a treatment, readers should consult *Minitab Handbook*, Second Edition (1985) by B.F. Ryan, B.L. Joiner and T.A. Ryan, Duxbury Press, Boston.

In order to use Minitab it must, naturally, be loaded into your computer. After that you need to learn a number of simple commands. If you make a mistake typing in a command, the resulting error message is usually self-explanatory. However, do not expect Minitab to tell you which statistical method you should be using – Minitab is not an expert system.

16.2 GETTING STARTED

The package is accessed after logging in by typing in the command **MINITAB**. Then press the RETURN[†] key after this command, and in general after the end of each line of a Minitab program. The screen response to the command **MINITAB** is **MTB>**. This implies that you have 'entered' the package and you may now type in further commands.

16.3 DATA INPUT

Data are held in a 'worksheet' consisting of rows and columns. Often the data for a particular variable is held in one column. The columns are referred to as C1, C2,. . . . If the command **READ C1** is typed in, the data must then be entered as one column of values. Alternatively, the command **SET C1** enables the values to be held in column 1 of the worksheet to be

[†] or whichever key is used to get your computer to accept a command.

entered in rows. (The main advantage of the **READ** command is that several variables can be read in at the same time by using **READ C1, C2,** . . ., whereas only one variable is allowed as the argument of the **SET** command.)

Example

Enter the data from Table 3.1 in column 1, and obtain a screen printout.

Table 16.1 An example of a Minitab program to read in and print out 50 values

MTB>	**SET C1**									
DATA>	**164.8**	**182.7**	**168.4**	**165.6**	**155.0**	**176.2**	**171.2**	**181.3**	**176.0**	**178.4**
DATA>	**169.0**	**161.4**	**176.8**	**166.1**	**159.9**	**184.0**	**171.8**	**169.2**	**163.7**	**160.2**
DATA>	**173.7**	**153.6**	**151.7**	**165.1**	**174.2**	**174.8**	**173.0**	**157.8**	**169.7**	**179.7**
DATA>	**172.2**	**174.6**	**179.6**	**146.3**	**169.3**	**164.5**	**175.0**	**193.9**	**184.6**	**163.6**
DATA>	**177.7**	**190.1**	**168.7**	**170.0**	**165.3**	**180.1**	**168.3**	**185.4**	**171.3**	**186.8**
DATA>	**END**									
MTB>	**PRINT C1**									

Notes

(a) The prompt **DATA>** appears in response to the command **SET C1**.
(b) Remember to press the RETURN key at the end of each line of data.
(c) The command **END** implies the end of the data.
(d) The data held in C1 will appear on the screen after the command **PRINT C1** has been entered (and the RETURN key has been pressed).

16.4 EDITING DATA

Suppose you make a mistake entering one or more data values. If you notice that the mistake is on the current line you can back-track by pressing the ← key. If the mistake is on a line already entered into the worksheet (i.e. you have already pressed the RETURN key at the end of the line) then carry on entering the rest of the data. Having completed this you can edit the data by one of four commands, namely **LET**, **INSERT**, **DELETE**, or **ERASE**. For example:

1. Suppose you typed in 166.5 instead of 165.6 (fourth value in C1). Then **MTB> LET C1(4) = 165.6** would make the correction.
2. Suppose you omitted 161.4 (twelfth value in C1). Then
   MTB> INSERT 11 12 C1
   DATA> 161.4
   DATA> END
   would insert 161.4 between 169.0 (the eleventh value entered) and 176.8 (the twelfth value entered).

3. Suppose you typed in 174.2 (twenty-fifth value in C1) twice instead of only once. Then
   MTB> DELETE 26 C1
   would delete the twenty-sixth value (174.2) in C1.
4. Suppose you decided to re-enter C1 from scratch, having made several mistakes. Then
   MTB> ERASE C1
   would erase all the values in C1.

When you have completed the editing, enter the command **PRINT C1** to check that all the values in C1 are now correct.

16.5 SUMMARIZING DATA BY GRAPHICAL METHODS AND NUMERICAL SUMMARIES

This was discussed in Chapters 3 and 4. If the Minitab commands in Table 16.1, ending in **PRINT C1**, are followed by the commands **NAME**, **DOTPLOT**, **HISTOGRAM** and **DESCRIBE** as detailed in Table 16.2, the display on the screen would be as shown in the table.

Table 16.2 An example of summarizing data

MTB> NAME C1 'HEIGHT'

MTB> DOTPLOT 'HEIGHT'

MTB> HISTOGRAM 'HEIGHT';

SUBC> INCREMENT 10;

SUBC> START 150.

Histogram of HEIGHT N = 50

Midpoint	Count	
150.0	3	***
160.0	9	*********
170.0	21	*********************
180.0	13	*************
190.0	4	****

MTB> DESCRIBE 'HEIGHT'

	N	MEAN	MEDIAN	TRMEAN	STDEV	SEMEAN
HEIGHT	50	171.25	171.25	171.36	9.91	1.40
	MIN	MAX	Q1	Q3		
HEIGHT	146.30	193.90	165.02	177.87		

Notes

(a) In the dotplot each data value is represented by a dot. This plot is, therefore, similar to the cross-diagram (see Fig. 3.9).

(b) Three commands were used to obtain the histogram. The semi-colon at the end of the **HISTOGRAM** command ensures that the sub-command prompt **SUBC>** will appear in the next line. The first sub-command indicates that the difference between the mid-points of adjacent groups is 10. This sub-command also ends with a semi-colon. The second sub-command indicates that the mid-point of the first group is 150. This is the last sub-command and must therefore end with a full-stop.

(c) The Minitab histogram is the same shape as the conventional histogram (see Fig. 3.1) except for a rotation of 90 degrees.

(d) The summary statistics following the **DESCRIBE** command apply to the data held in C1:

N is the number of data values held in C1
MEAN is the mean of the ungrouped data (Section 4.2)
MEDIAN is the median of the ungrouped data (Section 4.3)
TRMEAN is the trimmed mean, i.e. the mean excluding the smallest 5% and largest 5% of the values.
STDEV is the standard deviation of the ungrouped data (Section 4.7)
SEMEAN is the standard error of the mean and equals **STDEV/$\sqrt{}$N**
MIN, **MAX** are the smallest and largest values, respectively.
Q1, **Q3** are the lower and upper quartiles, respectively, of the ungrouped data (Section 4.8).

16.6 SAVING DATA FOR A FUTURE MINITAB SESSION, AND LEAVING MINITAB

Suppose you wish to finish a Minitab session but you also wish to return later to continue analysis of the data held in the worksheet. This may be achieved by the **SAVE**, **STOP** and **RETRIEVE** commands. For example, if the commands following **DESCRIBE 'HEIGHT'** in Table 16.2 are

MTB> SAVE 'HEIGHTDAT'
MTB> STOP

then the 50 height values held in C1 will be saved in a Minitab file called **HEIGHTDAT**. Notice that in the **SAVE** command, the name of the file must be enclosed in single quotes.

The command **STOP** will take you out of Minitab and return you to the operating system of your computer.

At a future Minitab session, in order to continue analysis of these data the required commands are

MINITAB (to 'enter' the package)
MTB> RETRIEVE 'HEIGHTDAT'

This will restore the worksheet to what it was when the **SAVE** command was typed in. The names of variables are also saved.

16.7 THE HELP COMMAND

Information about Minitab is stored in the package. To get information on any commands, e.g. **SET**, simply typed **HELP SET**. To get information about other available help, type **HELP HELP**.

16.8 GETTING HARD-COPY PRINTOUTS

At some stage you will probably wish to obtain a printout on paper of a Minitab session. One way of achieving this is to employ the Minitab commands **OUTFILE** and **NOOUTFILE**, together with a command such as **SPOOL** when you have left Minitab and returned to your computer's operating system. Table 16.3 shows the correct positioning of these commands as they apply to the example started in Table 16.1, continued in Table 16.2, and assuming that you wish to save the data.

Table 16.3 An example of how to obtain a hard-copy printout of a Minitab session

MTB>	**OUTFILE 'MINITAB01'**
MTB>	**SET C1**
DATA>	**164.8** etc.
MTB>	**PRINT C1**
MTB>	**NAME C1 'HEIGHT'**
MTB>	**DESCRIBE 'HEIGHT'**
MTB>	**SAVE 'HEIGHTDAT'**
MTB>	**NOOUTFILE**
MTB>	**STOP**
OK, SPOOL MINITAB01.LIS	

Notes

(a) The printout will contain all the Minitab commands and responses between the **OUTFILE** and **NOOUTFILE** commands, except for the data (see lines beginning **DATA>**). Because of this exception it is a good idea to include a **PRINT** command immediately after the data have been read into the worksheet.

(b) Note that in the **OUTFILE** command the name of the file, **MINITAB01**, must in enclosed in single quotes. This file is stored in the computer as **MINITAB01.LIS** (with no quotes), assuming Version 6.1.1 of Minitab is being used. (In an earlier version, namely 5.1.3, **.LIST** was required instead of **.LIS**). This note refers to running Minitab on a PRIME computer, and may not be applicable to other makes of computer.

(c) You should distinguish between Minitab files, such as **HEIGHTDAT**, and computer files, such as **MINITAB01.LIS**. Minitab files cannot be edited by the computer's editor.

16.9 DISCRETE PROBABILITY DISTRIBUTIONS

Minitab can be used to generate probabilities for both the binomial and Poisson distributions (which were introduced in Chapter 6). The **PDF** command produces probabilities of exact numbers of successes for a specified binomial distribution or exact numbers of random events in time or space for a specified Poisson distribution. The **CDF** command produces cumulative probabilities of so many or fewer successes (binomial) or random events (Poisson). Examples are given in Tables 16.4 and 16.5.

Table 16.4 Probabilities for the $B(10, 0.5)$ distribution

MTB> PDF;

SUBC> BINOMIAL N = 10, P = 0.5.

BINOMIAL WITH	**N = 10 P = 0.5**
K	**P(X = K)**
0	**0.0010**
1	**0.0098**
2	**0.0439**
3	**0.1172**
4	**0.2051**
5	**0.2461**
6	**0.2051**
7	**0.1172**
8	**0.0439**
9	**0.0098**
10	**0.0010**

Notice that the probabilities in Table 16.4 are equal to those in Table 6.2, apart from rounding. Also note that the probabilities in Table 16.5 are identical with those in Table D.1 for $n = 10$, $p = 0.5$.

Poisson probabilities can be obtained in the same way. Try:

Table 16.5 Cumulative probabilities for the $B(10, 0.5)$ distribution

MTB> CDF;

SUBC> BINOMIAL N = 10, P = 0.5.

BINOMIAL WITH	N = 10 P = 0.5
K	P(X LESS or = K)
0	0.0010
1	0.0107
2	0.0547
3	0.1719
4	0.3770
5	0.6230
6	0.8281
7	0.9453
8	0.9893
9	0.9990
10	1.0000

MTB> PDF;
SUBC> POISSON MEAN = 5.

and compare with Table 6.3. Also try:

MTB> CDF;
SUBC> POISSON MEAN = 5.

and compare with Table D.2 for $m = 5$.

It is also possible to get Minitab to simulate binomial and Poisson experiments. For example, the $B(10, 0.5)$ distribution is a model for the

Table 16.6 100 simulations of a $B(10, 0.5)$ experiment

MTB> RANDOM 100 C1;

SUBC> BINOMIAL N = 10, P = 0.5.

MTB> PRINT C1

C1

3	6	6	6	8	6	9	6	4	6	6	6	6	6	6
6	6	5	6	5	8	3	6	7	3	3	7	5	6	6
5	7	6	7	3	4	6	4	6	6	3	6	8	5	4
6	5	5	8	4	4	5	6	3	7	2	5	3	6	6
5	3	3	8	5	2	4	5	5	5	2	4	7	6	7
5	3	5	3	5	5	7	5	5	6	6	6	5	3	5
4	5	4	7	5	5	4	6	8	6					

binomial experiment of tossing a coin 10 times and counting the number of heads. This experiment may be simulated 100 times, say, by the commands shown in Table 16.6.

If this experiment is carried out a large number of times, theory (see Section 6.5) suggests that the mean and standard deviation of the number of successes should be $np = 5$ and $\sqrt{np(1-p)} = 1.58$. Even with only 100 repetitions the **DESCRIBE C1** command (not shown in Table 16.6) gives a mean of 5.26 and standard deviation of 1.495, quite close to their 'theoretical' values. We could also compare the shape of the above data (using **HISTOGRAM C1**) with the theoretical shape (Fig. 6.1).

Poisson experiments may also be simulated. For example, if we know that the mean number of randomly arriving telephone calls at a switchboard is 5 calls per 5 minutes, we may simulate 100 periods each of 5 minutes and print the results with the commands:

```
MTB> RANDOM 100 C1;
SUBC> POISSON MEAN = 5.
MTB> PRINT C1
```

The theoretical values for the mean and standard deviation of the number of calls are $m = 5$ and $\sqrt{m} = 2.236$ (see Section 6.11). The agreement between the values obtained from the simulation and the theoretical values should get better as the number of repetitions increases.

16.10 CONTINUOUS PROBABILITY DISTRIBUTIONS

Minitab can be used to generate probabilities for a number of continuous probability distributions. Only the normal distribution (introduced in Chapter 7) will be discussed here. The **CDF** command gives the area to the left of a particular value of the variable for a specified normal distribution, giving probabilities identical to those of Table D.3(a). The **INVCDF** command gives the value of the variable for a particular area to the left of it, again for a specified normal distribution. An example is given in Table 16.7.

Table 16.7 The **CDF** and **INVCDF** commands for a $N(170, 10^2)$ distribution

```
MTB>  CDF 185;
SUBC> NORMAL MU = 170, SIGMA = 10.
    185.000 0.9332;
MTB>  INVCDF 0.9332;
SUBC> NORMAL MU = 170, SIGMA = 10.
    0.9332   185.0005
```

Notes
(a) Minitab gives an area of 0.9332 to the left of 185 in response to the **CDF** command and sub-command (see Fig. 7.5 for this distribution).
(b) The **INVCDF** command is the same calculation in reverse.

It is also possible to get Minitab to simulate values taken from a specified normal distribution. The commands in Table 16.8 will generate 100 random observations from the $N(170, 10^2)$ distribution which should have a mean close to 170 and a standard deviation close to 10. Try these commands.

Table 16.8 Simulation of 100 random observations from a $N(170, 10^2)$ distribution

```
MTB>  RANDOM 100 C1;
SUBC> NORMAL MU = 170, SIGMA = 10.
MTB>  PRINT C1
MTB>  DESCRIBE C1
```

16.11 SIMULATION OF THE SAMPLING DISTRIBUTION OF THE SAMPLE MEAN

In Section 8.5 we quoted that, if samples of size n are taken from a population with mean μ and standard deviation σ, the distribution of the sample mean:

1. has a mean μ;
2. has a standard deviation $\sigma/\sqrt{n}$;
3. is approximately normal if n is large, irrespective of the shape of distribution of the parent population. (However, if the 'parent' distribution is normal, the distribution of the sample mean is normal for all values of n.)

We can decide if the above theory seems reasonable by simulating samples from a known parent population and looking at aspects of the distribution of the sample mean. An example is given in Table 16.9.

The **RANDOM** command and the sub-command put 200 randomly selected values from a normal distribution with a mean of 100 and a standard deviation of 15 into each of columns 1 to 9. We therefore have 200 rows of values in our worksheet, each row consisting of nine values. The **RMEAN** command calculates the mean of each row, i.e. sample, of size 9 and puts the answer in column 10. Column 10 is thus a simulation of 200 values from the distribution of the sample mean, for samples of size 9. The theoretical values for the mean and standard deviation of this distribution

Table 16.9 Distribution of sample of size 9 taken from a $N(170, 15^2)$ distribution

```
MTB> RANDOM 200 C1-C9;
SUBC> NORMAL MU = 100, SIGMA = 15.
MTB> RMEAN C1-C9 INTO C10
MTB> PRINT C10
MTB> DESCRIBE C10
MTB> HISTOGRAM C10
```

are 100 and $15/\sqrt{9} = 5$, and we can see how close we get to these by the output following the **DESCRIBE** command. We can also check visually whether the sampling distribution seems normal by using the **HISTOGRAM** command.

The reader is urged to try

1. varying the number of samples selected (200 in the example) for the same 'parent';
2. varying the sample size (9 in the example) for the same 'parent';
3. a non-normal parent, for example a Poisson distribution with a mean of 0.5.

16.12 CONFIDENCE INTERVAL ESTIMATION FOR MEANS

In this section we illustrate the Minitab commands required to obtain confidence intervals for three of the cases considered in Chapter 9, namely those discussed in Sections 9.4, 9.10, 9.11. Naturally all the assumptions required in using the various formulae still apply.

Table 16.10 A 95% confidence interval for μ

```
MTB> SET C1
DATA> 1.02 1.05 1.08 1.03 1.00 1.06 1.08 1.01 1.04 1.07 1.00
DATA> END
MTB> TINTERVAL 95 C1
      N    MEAN      STDEV     SEMEAN    95.0 PERCENT C.I.
C1   11    1.04000   0.03033   0.00915   (1.01962, 1.06038)
```

The data here are from Worksheet 10, Question 7. The output following the **TINTERVAL** command shows that a 95% confidence interval for the mean weight for the batch (population) is 1.02 to 1.06 (2 dp). The formula used is that shown in Section 9.4.

For paired samples data (Section 9.10), the differences between pairs are calculated using the **LET** command as in Table 16.11, which uses the test scores from Table 9.1. The output following the **TINTERVAL** command gives a confidence interval for μ_d of 1.44 to 13.56 (agreeing with values stated in Section 9.10).

Table 16.11 A 95% confidence interval for the mean of a population of differences

```
MTB>  SET C1
DATA> 56    59    61    48    39    56    75    45    81    60
DATA> END
MTB>  SET C2
DATA> 63    57    67    52    61    71    70    46    93    75
DATA> END
MTB>  LET C3 = C2 - C1
MTB>  NAME C3 'DIFF'
MTB>  TINTERVAL  95  'DIFF'
        N    MEAN    STDEV    SEMEAN      95.0    PERCENT C.I.
DIFF   10    7.5     8.48     2.68       (1.44,   13.56)
```

For unpaired samples data (Section 9.11) the following commands will result in a 95% confidence interval for μ_1 and μ_2, assuming that the ungrouped data from the two samples have been read into columns 1 and 2:

```
MTB> TWOSAMPLE-T 95  C1  C2;
SUBC> POOLED.
```

The sub-command **POOLED** ensures that a pooled estimate of the common variance is used in the calculation of the confidence interval. (A numerical example of the **TWOSAMPLE-T** command is given in Section 16.14, since the same command is also employed when an unpaired t-test is required.)

16.13 SIMULATION AND CONFIDENCE INTERVALS

Confidence intervals are relatively simple to calculate, but what do they mean when we have calculated them? In the case of 95% confidence intervals for a population mean, μ, the answer is given by the last two sentences of Section 9.2. (Similar statements can be made for confidence levels other than 95% and for other parameters such as μ_d and $\mu_1 - \mu_2$. To illustrate the concept involved we can use simulation to take a number of samples of the same size, n, from a normal distribution with a known mean μ and standard deviation σ. For each sample we then calculate a

95% confidence interval. If the formula used to calculate the 95% confidence intervals is correct, then we would expect that 95% of such confidence intervals to 'capture' the mean μ. By 'capture' we mean that the known value of μ lies inside the confidence interval.

Table 16.12 100 Confidence intervals from 100 samples of size 9 taken from a $N(70, 3^2)$ distribution

```
MTB>  RANDOM 9   C1-C50;
SUBC> NORMAL MU = 70   SIGMA = 3.
MTB>  RANDOM 9   C51-C100;
SUBC> NORMAL MU = 70   SIGMA = 3.
MTB>  TINTERVAL   95   C1-C100
```

Table 16.12 is an example of such a simulation. It will produce a total of 100 confidence intervals for μ, which we already know is 70. We can count how many intervals contain the value 70 and compare with the theory which indicates that 95 of the 100 intervals are expected to do so.

16.14 HYPOTHESIS TESTING FOR MEANS

In this section we illustrate the Minitab commands required to test hypotheses for three of the cases considered in Chapter 10, namely those discussed in Sections 10.9, 10.12 and 10.13. Naturally all the assumptions required in using the various formulae still apply.

Table 16.13 Hypothesis test for μ

```
MTB>  SET C1
DATA> 1.02  1.05  1.08  1.03  1.00  1.06  1.08  1.01  1.04  1.07  1.00
DATA> END
MTB>  TTEST  MU = 1  C1
TEST  OF  MU = 1.00  VS  MU  N.E.    1.00

      N    MEAN      STDEV     SEMEAN     T      P VALUE
C1    11   1.04000   0.03033   0.00915    4.37   0.0014
```

The data in Table 16.13 are from Worksheet 10, Question 7. The null hypothesis is $H_0 : \mu = 1$ and the alternative hypothesis is $H_1 : \mu \neq 1$. If we wished to have a different alternative hypothesis, then the subcommand:

SUBC> ALTERNATIVE = +1. would be needed for $H_1 : \mu > 1$
while
SUBC> ALTERNATIVE = −1. would be needed for $H_1 : \mu < 1$.

Note that in both these cases the **TTEST** command would need to end with a semi-colon.

The output gives *Calc* $t = 4.37$ but, unlike the t-test discussed in Chapter 10, does not give *Tab* t. Instead a p-value is given. A p-value of less than 0.05 implies that we should reject H_0 at the 5% level of significance.

For paired samples data (Section 10.12) the differences between pairs are calculated using the **LET** command. Table 16.14 shows a paired samples t-test for the data from Table 9.1, where the null hypothesis $H_0: \mu_d = 0$ is tested against the alternative $H_1: \mu_d > 0$. Since the p-value is less than 0.05, H_0 is rejected at the 5% level of significance (agreeing with the conclusion of Section 10.12).

For the unpaired samples t-test (Section 10.13) the command and sub-command given at the end of Section 16.12 not only provide a confidence

Table 16.14 A paired samples t-test

```
MTB>  SET C1
DATA> 56     59     61     48     39     56     75     45     81     60
DATA> END
MTB>  SET C2
DATA> 63     57     67     52     61     71     70     46     93     75
DATA> END
MTB>  LET C3 = C2 - C1
MTB>  NAME C3 'DIFF'
MTB>  TTEST  0  'DIFF';
SUBC> ALTERNATIVE +1.
        N    MEAN     STDEV     SE MEAN     T      P VALUE
DIFF    10   7.50     8.48      2.68        2.80   0.010
```

Table 16.15 An unpaired samples t-test

```
MTB>  SET    C1
DATA> 4600   4710   4820   4670   4760   4480
DATA> END
MTB>  SET    C2
DATA> 4400   4450   4700   4400   4170   4100
DATA> END
MTB>  TWOSAMPLE-T 95  C1  C2;
SUBC> POOLED.
TWOSAMPLE   T  FOR  C1  VS  C2
   N        MEAN        STDEV   SEMEAN
C1  6       4673        121     49
C2  6       4370        214     88
95  PCT  CI   FOR MU C1 - MU C2: (79, 527)

TTEST    MU   C1 = MU C2   (VS NE): T = 3.02  P = 0.013  DF = 10.0
```

interval for $\mu_1 - \mu_2$ but can also be used to test the null hypothesis $H_0: \mu_1 - \mu_2$ against an alternative, for example $H_1: \mu_1 \neq \mu_2$. Once again *Calc t* and a *p*-value are output. The example in Table 16.15 refers to Worksheet 10, Question 16. One-sided alternative hypotheses may be included by adding an **ALTERNATIVE** sub-command after the **TWOSAMPLE-T** command.

16.15 NON-PARAMETRIC HYPOTHESIS TESTS

In Chapter 11 three non-parametric hypothesis tests were described, namely the sign test, the Wilcoxon signed rank test and the Mann–Whitney *U* test. Table 16.16 shows an example of a sign test for the data from the example in Section 11.3. The *p*-value of 0.0547 indicates that H_0 should not be rejected at the 5% level of significance.

Table 16.16 A sign test for the median of a population of differences

```
MTB>  SET  C1
DATA> 7    -2    6    4    22    15    -5    1    12    15
DATA> END
MTB>  NAME  C1  'DIFF'
MTB>  STEST      0  'DIFF';
SUBC> ALTERNATIVE +1.
SIGN  TEST     OF  MEDIAN = 0.00  VERSUS  MEDIAN  G.T.  0.00
      N    BELOW   EQUAL      ABOVE    P-VALUE  MEDIAN
DIFF  10   2       0          8        0.0547     6.50
```

The small-sample Wilcoxon signed rank test, described in Section 11.5, is not available from Minitab. Instead, the large-sample test, described in Section 11.6, is employed. As an example we can simply replace the **STEST** command in Table 16.16 by the **WTEST** command. The output gives a *p*-value.

The small-sample Mann–Whitney *U* test, described in Section 11.7, is not available from Minitab. Instead, the large-sample test, described in Section 11.8, is employed. The output indicates the level at which the test of the null hypothesis that the medians are equal, against a two-sided alternative, is significant. (Minitab uses **ETA1** and **ETA2** to denote the medians.)

16.16 χ^2 TEST FOR INDEPENDENCE, CONTINGENCY TABLE DATA

In this section we refer to the tests described in Chapter 12. Table 16.17 shows how Minitab performs a χ^2 test for the data from Table 12.2.

Table 16.17 A χ^2 test of independence for a 2 × 2 contingency table

```
MTB>  READ C1   C2
DATA> 30   10
DATA> 20   40
DATA> END
MTB>  CHISQUARE   C1   C2
Expected counts are printed below observed counts
           C1       C2       Total
      1    30       10       40
           20.0     20.0
      2    20       40       60
           30.0     30.0
Total      50       50       100
Chisq      5.00  +  5.00  +
           3.33  +  3.33  =  16.67
df = 1
```

Note that Minitab does not use Yates's continuity correction for 2 × 2 contingency tables, so the χ^2 value of 16.67 does not agree with the value of 15.04 obtained in Section 12.3.

It was noted in Section 12.4 that the formula for *Calc* χ^2 is theoretically valid only if all E values are sufficiently large. Minitab handles this problem by giving the number of cells in the contingency table with expected values of less than 5. This leaves open the option of combining rows or columns and re-entering the combined observed frequencies. For example, if the observed frequencies from Table 12.4 are entered in a Minitab program to perform a χ^2 test, the message

3 cells with expected frequencies less than 5

is printed. Combining rows 1 and 2 overcomes the problem; see Table 12.5.

16.17 SCATTER DIAGRAMS AND CORRELATION

In Chapter 13 two correlation coefficients were discussed, namely Pearson's r and Spearman's r_s. The latter is not available in Minitab, but Table 16.18 shows how to obtain a scatter diagram (using the **PLOT** command) and Pearson's r for the data in Table 13.1.

The first variable in the **PLOT** command is plotted on the vertical axis, the second on the horizontal axis. The value of the correlation coefficient is output, but the hypothesis test for ϱ (see Section 13.3) is not available in Minitab.

Table 16.18 A scatter diagram and Pearson's *r*

```
MTB>  SET  C1
DATA> 170  175  176  178  183  185
DATA> END
MTB>  SET  C2
DATA> 57  64  70  76  71  82
DATA> END
MTB>  NAME  C1  'HEIGHT'
MTB>  NAME  C2  'WEIGHT'
MTB>  PLOT  'HEIGHT'  'WEIGHT'
MTB>  CORRELATION  'HEIGHT'  'WEIGHT'
```

16.8 REGRESSION ANALYSIS

In Chapter 14 we discussed only simple linear regression analysis (see Section 14.1). Table 16.19 illustrates some of the regression facilities available in Minitab. The data are from Table 14.1.

Table 16.19 Simple linear regression analysis

```
MTB>  SET  C1
DATA> 105  120  160  155  70  150  185  130
DATA> END
MTB>  SET  C2
DATA> 45  75  85  65  50  70  80  55
DATA> END
MTB>  NAME  C1  'SALES'
MTB>  NAME  C2  'SCORE'
MTB>  BRIEF  1
MTB>  REGRESSION  'SALES'  1  'SCORE';
SUBC> PREDICT  60.
The regression equation is
SALES = 9.4 + 1.90 SCORE
Fit           95% C.I.
123.66        (100.14, 147.19)
```

The command **BRIEF 1** restricts the amount of output (even so, the output shown in Table 16.19 is only part of Minitab's output). The regression equation agrees with that obtained in Section 14.2, the 'fit' is the predicted value of sales for a score of 60 (see Section 14.4) and the 95% C.I. is the 95% confidence interval for predicted sales for a score of 60 (see Section 14.5). In the **REGRESSION** command the **1** (between **'SALES'** and **'SCORE'**) indicates that there will be only one *x* variable, namely **'SCORE'**.

16.19 χ^2 GOODNESS-OF-FIT TESTS

Four χ^2 goodness-of-fit tests were discussed in Chapter 15. However, there are no standard Minitab commands to perform these tests. The best we can do is to write a Minitab program to output expected frequencies and *Calc* χ^2, and hope that the expected frequencies are not too small. Table 16.20 and Table 16.21 show examples of programs relating to the examples of goodness-of-fit tests for the simple proportion distribution described in Section 15.2 and the binomial distribution described in Section 15.3, respectively.

Table 16.20 Expected values and *Calc* χ^2 for the goodness-of-fit test in Section 15.2

```
MTB>  SET  C1
DATA> 5  3  1  1
DATA> END
MTB>  SET  C2
DATA> 28  12  6  4
DATA> END
MTB>  LET  C3 = C1/SUM(C1)
MTB>  LET  C4 = C3*SUM(C2)
MTB>  PRINT  C4
C4
        25  15  5  5
MTB>  LET  K1 = SUM((C2-C4)**2/C4)
MTB>  PRINT  K1
K1      1.36
```

Notes

(a) The theoretical ratios and observed frequencies are read into C1 and C2, respectively.

(b) Expected proportions and frequencies are put into C3 and C4, respectively. Note that all the expected values are at least 5.

(c) The value of *Calc* χ^2 is given by the value of K1. The test may be completed by steps 5, 6 and 7 as in Section 15.2.

Table 16.21 Expected values for the goodness-of-fit test in Section 15.3

```
MTB>  SET  C1
DATA> 0  1  2  3  4  5
DATA> END
MTB>  SET  C2
DATA> 15  18  8  5  3  1
DATA> END
MTB>  PDF  C1  C3;
```

```
SUBC> BINOMIAL  N=5, P=0.25.
MTB> LET  C4 = C3*SUM(C2)
MTB> PRINT  C4
C4
     11.8652  19.7754  13.1836  4.3945  0.7324  0.0488
```

Notes

(a) The values of the variable, x, and the observed frequencies are read into C1 and C2, respectively.

(b) Binomial probabilities are put into C3.

(c) Expected frequencies are put into C4. Since some of these are less than 5, they should not be used as they stand in the calculation of χ^2. It is necessary to combine the last three categories (as in Table 15.3) and then proceed to obtain *Calc* χ^2 and complete the test.

16.20 SUMMARY

A substantial proportion of the statistical procedures described in this book may be performed on a computer using the interactive statistical computer package, Minitab. In addition, Minitab is particularly valuable in simulation work, which helps us to understand, for example, the concepts of sampling distributions and confidence intervals.

No special knowledge of computing is required to make use of Minitab other than the ability to log in, type in simple commands and respond to screen messages. However, it is the user's sole responsibility to choose the correct statistical procedure each time a statistical problem is tackled. The facility for rapid calculations and the ability to generate pages and pages of computer output are not a substitute for correct statistical thinking.

Table 16.22 gives an alphabetical index of the Minitab commands and sub-commands illustrated in this chapter.

Table 16.22 Minitab commands

Command	*Use*	*Section reference of Chapter 16*
BRIEF 1	Limiting regression output	18
CDF	Probabilities for distributions	9, 10
CHISQUARE	χ^2 test	16
CORRELATION	Pearson's correlation coefficient	17
DELETE	Deleting a row of data	4
DESCRIBE	Summary statistics	5, 8, 10, 11
DOTPLOT	Horizontal plot for one variable	5
END	End of a data list	3, 11, 12, 14, etc.
ERASE	Erasing a column of data	4

Table 16.22 Minitab commands (contined)

Command	*Use*	*Section reference of Chapter 16*
HELP	Getting help with Minitab	7
HISTOGRAM	Histogram for one variable	5, 11
INSERT	Inserting a row of data	4
INVCDF	Reverse of **CDF**, for normal distribution	10
LET	Editing individual data values,	4
	creating new variables	12, 14, 15, 19
MANNWHITNEY	Mann–Whitney *U* test	15
MINITAB	To 'enter' Minitab, i.e. to start a Minitab session	2
NAME	Giving a name to a column variable	5, 8, 12, 14, etc.
NOOUTFILE	Ending an output file (for hard copy)	8
OUTFILE	Creating an output file (for hard copy)	8
PDF	Binomial and Poisson probabilities	9, 19
PLOT	Scatter diagrams	17
PRINT	Printing data and output on screen	3, 8, 10, 19
RANDOM	Generating random numbers	9, 10, 11, 13
READ	Inputting data in columns to be stored in the same columns	3, 16
REGRESSION	Regression analysis	18
RETRIEVE	Retrieving a Minitab worksheet	6
RMEAN	Calculating the mean of a row of data	11
SAVE	Saving a Minitab worksheet	6, 8
SET	Inputting data for one column variable in row	3, 8, 12, 14, etc.
STEST	Sign test	15
STOP	To 'leave' Minitab, i.e. to end a Minitab session	6, 8
TINTERVAL	Confidence interval for μ, μ_d	12, 13
TTEST	Hypothesis tests for μ, μ_d	14
TWOSAMPLE-T	Confidence interval and hypothesis test for $(\mu_1 - \mu_2)$	12, 14
WTEST	Wilcoxon signed rank test	15
Sub-command	*Used with command*	
ALTERNATIVE	**TTEST, STEST, WTEST**	
BINOMIAL	**CDF, PDF, RANDOM**	
INCREMENT	**HISTOGRAM**	
NORMAL	**CDF, INVCDF, RANDOM**	
POISSON	**CDF, PDF, RANDOM**	
POOLED	**TWOSAMPLE-T**	
PREDICT	**REGRESSION**	
START	**HISTOGRAM**	

WORKSHEET 16: MINITAB

1. (a) For the data given in Worksheet 3, Question 2(b), read the data into C1 using the **SET** command. If necessary edit the data. Obtain a histogram with mid-points at 0, 1, and so on. Also find the mean, standard deviation, median and inter-quartile range of the number of calls using the **DESCRIBE** command.
   (b) Obtain suitable graphical and numerical summaries for the data in
      (i) Worksheet, Question 2(d);
      (ii) Worksheet 3, Question 2(a).

2. Use the **PDF** command with the appropriate distributions to obtain the probabilities required in Worksheet 6, Questions 6, 9, 16 and 17.

3. Use the **CDF** command with the appropriate distributions to obtain the probabilities required in Worksheet 6, Questions 8 and 19.

4. In a multiple choice test there are five possible answers to each of 20 questions. Carry out a simulation for 100 candidates assuming that each candidate guesses the answer to each question and print out the results on the screen. Do the mean and standard deviation of the number of correct answers (out of a maximum of 20) agree with their theoretical values?

5. Use the **CDF** command with the appropriate distributions to obtain the probabilities required in Worksheet 7, Questions 4 and 10.

6. Use the **INVCDF** command with the appropriate distribution to obtain the answer to Worksheet 7, Question 11.

7. Simulate the weights of 50 oranges having a normal distribution with mean 70 g and standard deviation 3 g, and print out their weights on the screen. Count how many weigh: (a) over 75 g, (b) under 60 g, (c) between 60 and 75 g. Compare your answers with the theoretical values of 2, 0, 48 (given in the solution to Worksheet 7, Question 4).

8. In Minitab, it is possible to simulate an arbitrary discrete probability distribution by entering the values and the corresponding probabilities, and then using the **RANDOM** command. So the experiment of throwing a die 1000 times may be simulated as follows, the 1000 scores being placed in C3.

```
MTB> READ C1 C2
DATA> 1 0.16667
DATA> 2 0.16667
DATA> 3 0.16667
DATA> 4 0.16667
DATA> 5 0.16667
```

```
DATA> 6  0.16667
DATA> END
MTB>  RANDOM  1000  C3;
SUBC> DISCRETE  C1  C2.
MTB>  PRINT  C3
```

Now write a program to simulate the following experiment:

> Throw two dice and note the mean of the two scores. Repeat a total of 1000 times. Use the **DESCRIBE** commands to estimate the mean and standard deviation of the sampling distribution (for samples of size 2). Compare with their theoretical values (see Worksheet 8, Question 4). Check the shape of the distribution using the **HISTOGRAM** command. Repeat the simulation for three dice, four dice, etc.

9. Answer Worksheet 9, Questions 12 and 13 using appropriate Minitab commands.
10. Answer Worksheet 10, Question 14 using appropriate Minitab commands.
11. Answer Worksheet 12, Questions 8, 9 and 10 using the **CHISQUARE** command.
12. Use the **PLOT** and **CORRELATION** commands on the data from Worksheet 13, Question 7.
13. Use the **PLOT** and **REGRESSION** commands and the **PREDICT** sub-command to answer Worksheet 14, Questions 7 and 9.
14. Obtain the expected frequencies and *Calc* χ^2 for Worksheet 15, Questions 1 and 6.

Appendix A

MULTIPLE CHOICE TEST

Answer the following 50 questions by writing a, b or c according to your choice for each question; if you do not know the answer write d. You may use a calculator and statistical tables. Allow up to $1\frac{1}{2}$hours.

If you want to see how well you scored, give yourself 2 marks for a correct answer, −1 mark for an incorrect answer and 0 marks for a 'don't know'. (The marking means that the maximum mark is 100, and the average mark for 'guessers' is 0.)

1. $(\Sigma x)^2/n$ means:
   (a) Sum the n values of x, divide by n, and square,
   (b) Square each of the n values of x, sum, and divide by n,
   (c) Sum the n values of x, square, and divide by n.

2. Σfx^2 means:
   (a) Sum the f values, sum the x values, and then multiply the answers,
   (b) Multiply the f values by the x^2 values, and then sum,
   (c) Multiply the f values by the x values, square, and then sum.

3. The histogram is a graphical method of summarizing data when the variable is:
   (a) Discrete,
   (b) Categorical,
   (c) Continuous.

4. The cumulative frequency polygon may be used to obtain the following measures:
   (a) Mean and median,
   (b) Median and inter-quartile range,
   (c) Mean and standard deviation.

5. The average which represents the value which is exceeded as often as not is the:
   (a) Mean, (b) Median, (c) Mode.

6. The mean of the numbers 6, 7 and 8 is 7. If each number is squared the mean becomes:
   (a) 49, (b) Greater than 49, (c) Less than 49.

7. The standard deviation of the numbers 6, 7 and 8 is 1. If 1 is added to each number the standard deviation becomes:
   (a) 1, (b) 2 (c) $\sqrt{2}$.

8. A symmetrical distribution always has:
   (a) A bell shape,
   (b) A mean and a median with the same value,
   (c) No extremely high or low values.

9. If data exhibit a markedly skew distribution the best measure of variation is the:
   (a) Coefficient of variation,
   (b) Variance,
   (c) Inter-quartile range.

10. For a unimodal distribution with positive skewness:
    (a) mean > mode > median,
    (b) mean > median > mode,
    (c) mean < median < mode.

11. $P(B|A)$ means:
    (a) The probability of B divided by the probability of A,
    (b) The probability of B given that A has occurred,
    (c) The probability of A given that B has occurred.

12. If $P(A) = P(A|B)$, the events A and B are:
    (a) Mutually exclusive,
    (b) Exhaustive,
    (c) Statistically independent.

13. If three coins are tossed, the probability of two heads is:
    (a) 3/8, (b) 2/3, (c) 1/8.

14. A bag contains six red balls, four blue balls and two yellow balls. If two balls are drawn out with replacement, the probability that both balls will be red is:
    (a) 0.25, (b) 0.227, (c) 0.208.

15. A bag contains six red balls, four blue balls and two yellow balls. If two balls are drawn out without replacement, the probability that the first ball is blue and the second ball is yellow is:
    (a) 2/36, (b)4/33, (c) 2/33.

16. If two events A and B are statistically independent, the occurrence of A implies that the probability of B occurring will be:
    (a) 0, (b) unchanged, (c) 1.

17. $\binom{6}{3}$ is equal to:
    (a) 2, (b) 15, (c) 20.

18. For a binomial distribution with $n = 10$, $p = 0.5$, the probability of 5 or more successes is:
    (a) 0.5, (b) 0.623, (c) 0.377.

19. In a binomial experiment with three trials, the variable can take one of:
(a) 4 values, (b) 3 values, (c) 2 values.

20. If the number of random events per minute has a Poisson distribution, the maximum number of events which can occur in any one minute is:
(a) Unlimited, (b) 60, (c) 1.

21. For a Poisson distribution with a mean $m = 2$, P(2) is equal to:
(a) 0.2707, (b) 0.5940, (c) 0.7293.

22. The area to the left of $(\mu + \sigma)$ for a normal distribution is approximately equal to:
(a) 0.16, (b) 0.84, (c) 0.34.

23. For a normal distribution with mean μ and standard deviation σ,
(a) Approximately 5% of values are outside the range $(\mu - 2\sigma)$ to $(\mu + 2\sigma)$.
(b) Approximately 5% of values are greater than $(\mu + 2\sigma)$.
(c) Approximately 5% of values are outside the range $(\mu - \sigma)$ to $(\mu + \sigma)$.

24. For a normal distribution with $\mu = 10$, $\sigma = 2$, the probability of a value greater than 9 is:
(a) 0.6915, (b) 0.3085, (c) 0.1915.

25. The distribution of the means of samples of size 4, taken from a population with a standard deviation σ, has a standard deviation of:
(a) $\sigma/4$, (b) $\sigma/2$, (c) σ.

26. A 95% confidence interval for the mean of a population is such that:
(a) It contains 95% of the values in the population,
(b) There is a 95% chance that it contains all the values in the population,
(c) There is a 95% chance that it contains the mean of the population.

27. A confidence interval will be widened if:
(a) The confidence level is increased and the sample-size is reduced,
(b) The confidence level is increased and the sample-size is increased,
(c) The confidence level is decreased and the sample-size is increased.

28. Conclusions drawn from sample data about populations are always subject to uncertainty because:
(a) Data are not reliable,
(b) Calculations are not accurate,
(c) Only part of the population data is availabe.

29. The significance level is the risk of:
(a) Rejecting H_0 when H_0 is correct,
(b) Rejecting H_0 when H_1 is correct,
(c) Rejecting H_1 when H_1 is correct.

30. An example of a two-sided alternative hypothesis is:
(a) $H_1 : \mu < 0$, (b) $H_1 : \mu > 0$, (c) $H_1 : \mu \neq 0$.

31. If the magnitude of calculated value of t is less than the tabulated value of t, and H_1 is two-sided, we should:
(a) Reject H_0, (b) Not reject H_0, (c) Accept H_1.

32. Rejecting a null hypothesis H_0:
 (a) Proves that H_0 is false,
 (b) Implies that H_0 is unlikely to be true,
 (c) Proves that H_0 is true.

33. In an unpaired samples t-test with sample sizes $n_1 = 10$ and $n_2 = 10$, the value of tabulated t should be obtained for:
 (a) 9 degrees of freedom,
 (b) 19 degrees of freedom,
 (c) 18 degrees of freedom.

34. In analysing the results of an experiment involving six paired samples, tabulated t should be obtained for:
 (a) 11 degrees of freedom,
 (b) 5 degrees of freedom,
 (c) 10 degrees of freedom.

35. In a t-test for the mean of a population with $H_0 : \mu = 10$, $H_1 : \mu > 10$, a 5% significance level, and $n = 8$, the tabulated value of t is:
 (a) 2.365, (b) −1.895, (c) 1.895.

36. The sign test is:
 (a) Less powerful than the Wilcoxon signed rank test,
 (b) More powerful than the paired samples t-test,
 (c) More powerful than the Wilcoxon signed rank test.

37. The non-parametric equivalent of an unpaired samples t-test is the:
 (a) Sign test,
 (b) Wilcoxon signed rank test,
 (c) Mann–Whitney U test.

38. In the Wilcoxon signed rank test the null hypothesis is rejected if:
 (a) $Calc\ T \leqslant Tab\ T$,
 (b) $Calc\ T > Tab\ T$,
 (c) $Calc\ T < Tab\ T$.

39. The Mann–Whitney U test is preferred to a t-test when:
 (a) Data are paired,
 (b) Sample sizes are small,
 (c) The assumption of normality is invalid.

40. In order to carryout a χ^2 test on data in a contingency table, the observed values in the table should be:
 (a) Frequencies,
 (b) All greater than or equal to 5,
 (c) Close to the expected values.

41. In applying a χ^2 test to data from a 3×4 contingency table, at the 5% level of significance, the value of tabulated χ^2 is:
 (a) 19.7, (b) 12.6, (c) 14.5.

42. A significantly high positive value of a Pearson's correlation coefficient between two variables implies:

(a) That as one variable increases, the other variable decreases,
(b) A definite causal relationship,
(c) A possible causal relationship.

43. A random sample of 12 pairs of values have a Spearman rank correlation coefficient of 0.54. We conclude that, for a 5% level of significance:
(a) H_0 should be rejected in favour of a one-sided H_1,
(b) H_0 should be rejected in favour of a two-sided H_1,
(c) H_0 should not be rejected in favour of a one-sided H_1.

44. If the Pearson correlation coefficient between two variables is calculated to be -0.9 for a sample size of 16, the value of t is calculated using:

(a) $-0.9\sqrt{\left(\frac{14}{1-0.9^2}\right)}$, (b) $0.9\sqrt{\left(\frac{14}{1-0.9^2}\right)}$, (c) $-0.9\sqrt{\left(\frac{14}{1+0.9^2}\right)}$.

45. In regression analysis the y variable is chosen:
(a) Arbitrarily,
(b) As the variable plotted on the vertical axis in the scatter diagram,
(c) As the variable to be predicted.

46. The purpose of simple linear regression analysis is to:
(a) Replace points on a scatter diagram by a straight line,
(b) Measure the degree to which two variables are linearly associated,
(c) Predict one variable from another variable.

47. In a plant species green flowers occur twice as often as red flowers and equally as often as yellow flowers. In a random sample of 100 the expected number of yellow flowers will be:
(a) 20, (b) 40, (c) 25.

48. The number of degrees of freedom lost in a χ^2 goodness-of-fit test because parameters may have to be estimated from sample data could be as many as two for a:
(a) Poisson distribution,
(b) Binomial distribution,
(c) Normal distribution.

49. One of the following is not a null hypothesis:
(a) No correlation in the population between height and IQ.
(b) Hair colour and eye colour in a population are not independent,
(c) $\mu_1 = \mu_2$.

50. The purpose of statistical inference is:
(a) To draw conclusions about populations from sample data,
(b) To draw conclusions about populations and then collect sample data to support the conclusions,
(c) To collect sample data and use them to formulate hypotheses about a population.

Appendix B

SOLUTIONS TO WORKSHEETS AND MULTIPLE CHOICE TEST

WORKSHEET 1 (Solutions)

Key: C = continuous, D = discrete, R = ranked, Cat = categorical.

1. C; 0–1000 hours; light-bulb.
2. D; 0, 1, 2,. . . , 1000; year.
3. C; £5–£50; hotel.
4. D; 0, 1, 2,. . . , 100; hotel.
5. Cat; professional, skilled,. . . ; adult male.
6. D; 0, 1,. . . , 10; 100 hours.
7. C; 0–100 hours; 100 hours.
8. D; 0, 1, 2,. . . , 1000; month.
9. R; 1, 2, 3,. . . , 20; annual contest.
10. C; 0–300 cm; county.
11. D; 0, 1,. . . , 20; year (in period 1900–83).
12. D; 0, 1,. . . , 10^6; oil rig (output in barrels).
13. D; 0, 1,. . . , 1000; 10-second period.
14. D; 0, 1,. . . , 10; series of 10 encounters with T-junction.
15. R; A, B, C, D, E, Fail; candidate.
16. Cat; black, brown,. . . ; person.
17. Cat; presence, absence; square metre of meadow.
18. C; 0–5 seconds; rat.
19. D; 0, 1, 2,. . . , 20; page.
20. C; 0–10 kg; tomato plant.

21. Cat; sand, limestone etc.; core sample.

22. C; 0–100%; sample.

23. Cat; Con., Lab., SDP, Lib.,...; person.

WORKSHEET 2 (Solutions)

The numerical answer to each question is given. For Questions 1, 3, and 4 the required sequence of operations for a typical scientific calculator is also given.

1. (a)	−1.8	1.3 + 2.6 − 5.7 =
(b)	3	10.0 − 3.4 − 2.6 − 1.0 =
(c)	33.58	2.3 × 14.6 =
(d)	0.000334	0.009 × 0.0273 × 1.36 =
		Note Display may be 3.34152 − 04, indicating that the decimal point should be moved four places to the left.
(e)	0.1575	2.3 ÷ 14.6 =
(f)	341.297	1 ÷ 0.00293 = *or* 0.00293 $1/x$
(g)	8.35	2.3 + 4.6 + 9.2 + 17.3 = ÷ 4 =
		or 2.3 M in 4.6 M + 9.2 M + 17.3 M + MR ÷ 4 =
		Note The second method uses the memory. Min puts a number into memory overwriting what is there, M + adds a number to the memory. MR recalls the number currently in memory and displays it.
(h)	5.29	28 x^y 0.5 =
(i)	0.125	0.5 x^y 3 =
(j)	0.0164	0.2 x^y 2 = M in 0.8 x^y 4 = × MR =
(k)	1	0.5 x^y 0 =
(l)	125	0.2 x^y 3+/− =
		Note +/− changes +3 to −3 here,
(m)	4.953	1.6 INV e^x
		Note Assuming e^x requires the prior use of the INV button.
(n)	0.2019	1.6 +/− INV e^x
(o)	0.839	10 × 24 = $\sqrt{}$ M in 13 ÷ MR =
(p)	8	0.5 +/− × 4 = M in 6 − MR =
(q)	24	4 INV $x!$
	1	1 INV $x!$
	720	6 INV $x!$
	1	0 INV $x!$
	not defined	3 +/− INV $x!$ gives error message on display.
	not defined	2.4 INV $x!$ also gives error message on display.

2. (a) 33.6
   (b) 0.00033
   (c) 0.16
   (d) 341.3
   (e) 300

3. (a) 55   1 M in 2 M + 3 M + ... 10 M + MR

(b) 3   1 M in 2 M + 3 M + ...5 M + MR ÷ 5 =
(c) 55   1 INV x^2 M in 2 INV x^2 M + ...5 INV x^2 M + MR
Note Assuming x^2 requires the prior use of the INV button.
(d) 44   1 × 2 = M in 3 × 4 = M + 5 × 6 = M + MR

4. 24   2 M in 3 M + ...4 M + MR
3   24 ÷ 8 =
576   24 INV x^2
84   2 INV x^2 M in 3 INV x^2 M + ...4 INV x^2 M + MR
0   2 − 3 = M in 3 − 3 = M + ...4 − 3 = M + MR
12   2 − 3 = INV x^2 M in 3 − 3 = INV x^2 M + ...4 −3 = INV x^2 M + MR
12   To calculate $84 - (24)^2/8$, use 24 INV x^2 ÷ 8 = M in 84 − MR =

5. 17.7, 3.54, 313.29, 85.27, 0, 22.612, 22.612

WORKSHEET 3 (Solutions)

1. (a) Group frequency distribution table plus histogram or cumulative frequency distribution table plus cumulative frequency polygon.
   (b) Group frequency distribution table plus bar chart or pie chart.
   (c) Simple list plus bar chart.
   (d) As for (a).
   (e) As for (a).
   (f) Group frequency distribution table with 11 groups, line chart.
   (g) As for (a).
   (h) Group frequency distribution table with 6 groups, line chart.
   (i) Group frequency distribution table with 6 groups, bar chart or pie chart.
   (j) Simple list plus cross-diagram.
   (k) Group frequency distribution table with 3 groups, bar chart or pie chart.
   (l) As for (a).

2. (a)

Weight (g)	*No. of jars*
200–200.99	13
201–201.99	27
202–202.99	18
203–203.99	10
204–204.99	1
205–205.99	1

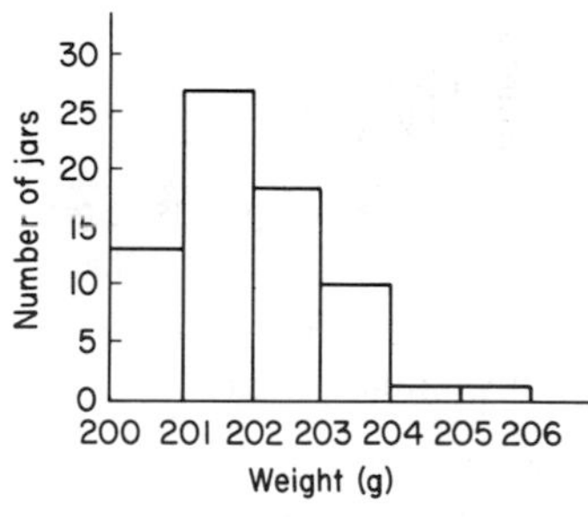

Fig. B.1 Histogram.

If two sets of 35 jars, draw-up two grouped frequency distribution tables, best way to compare graphically is to draw two cumulative frequency polygons.

(b) *No. of calls*	*No. of minutes*
0	21
1	13
2	8
3	3
4	4
5	1

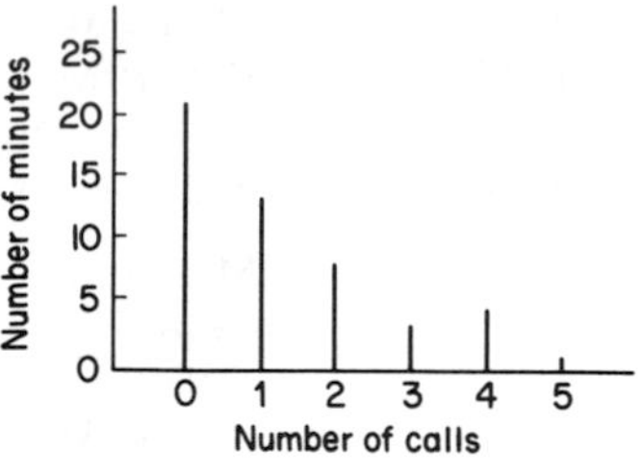

Fig. B.2 Line chart.

(i) Line chart.
(ii) Plot time series, number of cells versus number of minutes (see Fig. B.3). General impression is that the number of calls increases with time.

(c) *Distance from home to work (miles)*	*Number of staff*
Less than 5 miles	60
Less than 10 miles	160
Less than 15 miles	235
Less than 20 miles	280
Less than 25 miles	295
Less than 30 miles	299
Less than 35 miles	300

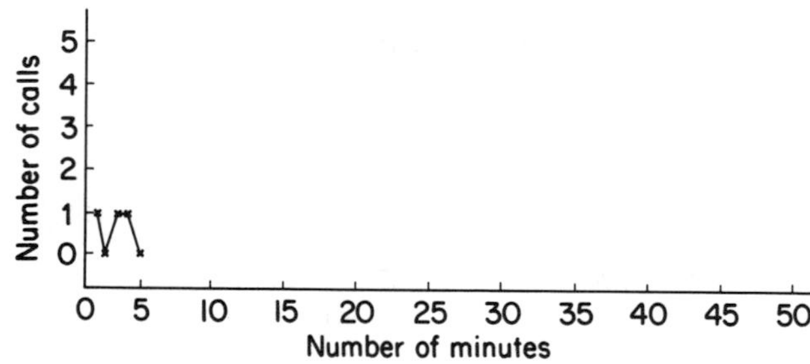

Fig. B.3 Time series graph of number of calls versus number of minutes (drawn for first five minutes only.)

Seems more accurate to say that either 1/3 of staff live less than 7 miles from work or 1/2 of staff live less than about $9\frac{1}{2}$ miles from work. (see Fig. B.4)

3. When rounding in the case of second decimal place values of 5, round first decimal place values down and up alternately. So 201·35 is rounded down to 201·3, 202·15 is rounded up to 202·2, and so on.

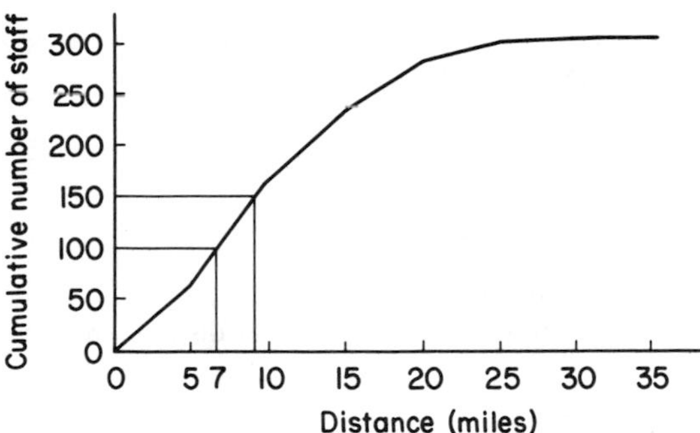

Fig. B.4 Cumulative frequency polygon.

(d) *A-level count*	*Number of students*
2	1
3	7
4	6
5	8
6	8
7	2
8	2
9	3
10	0
11	2
12	2
13	0
14	0
15	0
16	1

(see Fig. B.5)

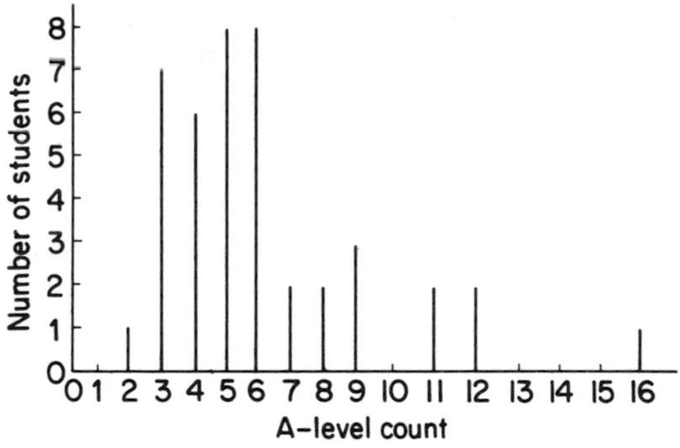

Fig. B.5 Line chart of A-level count. If IQ was also known, draw a scatter diagram.

	200	2 3 4 5 6 6 7 7 7 8 8 9 9
	201	1 1 2 2 3 3 3 3 4 4 4 5 6 6 6 6 6 7 7 7 7 8 8 8 9 9
	202	0 1 1 2 2 2 2 3 3 3 5 7 7 7 8 8 8 9 9
	203	0 1 1 2 3 4 5 5 7
	204	0 6
	205	7
$n = 70$	200	2 represents 200·2 gm

Shape is almost identical to histogram.
First two levels of stretched stem and leaf display are

200·	2 3 4
200*	5 6 6 7 7 7 8 8 9 9

WORKSHEET 4 (Solutions)

1. (a) They measure a middle value for a set of data.
   (b) Sample mode.
   (c) Sample mean.
   (d) Sample median.
   (e) Markedly skew data.
   (f) Only for categorical data (for which mean is not defined).
   (g) Roughly symmetrical data.
2. (a) Because averages can be misleading, and to show how much the data vary.
   (b) (i) Sample standard deviation, (ii) Sample inter-quartile range.
   (c) Question 3 below.
   (d) (i) Sample standard deviation, (ii) Sample inter-quartile range, (iii) None.
   (e) (i) Sample mean, (ii) Sample median, (iii) None.
3. (a) $\bar{x} = 78.1$, sample median = 67, sample mode = 62 or 67. Answers differ because one extremely high value, so positive skewness, and hence mean > median. Mode is non-unique.
   (b)

$$s = \sqrt{\left(\frac{89\,797 - \dfrac{937^2}{12}}{11}\right)} = 38.9.$$

   Sample inter-quartile range = 73.25 − 62 = 11.25.
   Answers differ because these two measures (standard deviation and inter-quartile range) measure variation in different ways.
   (c) Sample median and sample inter-quartile range, because of skewness in data. Coefficient of skewness = 3(78.1 − 67)/38.9 = 0.856.
   Alternatively, summarize by stating that one value was £200 and the other eleven are summarized by $\bar{x} = 67.0$, $s = 6.47$.
4.

$$\bar{x} = \frac{11.44}{11} = 1.04, \quad s = \sqrt{\left(\frac{11.9068 - \dfrac{11.44^2}{11}}{10}\right)} = 0.0303.$$

5. (a)

x	f	fx	fx^2
200.5	13	2606.5	522603.25
201.5	27	5440.5	1096260.75
202.5	18	3645.0	738112.50
203.5	10	2035.0	414122.50
204.5	1	204.5	41820.25
205.5	1	205.5	42230.25
	$\Sigma f = 70$	$\Sigma fx = 14137.0$	$\Sigma fx^2 = 2855149.50$

$\bar{x} = 201.96$, $s = 1.086$.

Note If your calculator has less than 9 figures on the display, simply subtract 200 (or better still 200.5) from each value of x. This has no effect on the value of s, but you will need to add 200 (or 200.5) to your answer for $\bar{x}$.

(b) From cumulative frequency polygon, median = 201.8, inter-quartile range = 1.5.

(c) Measure of skewness = 3(201.96 − 201.8)/1.086 = 0.442, not great so prefer $\bar{x}$ and s.

6. (a)

x	f	fx	fx^2
20	5	100	2000
28	10	280	7840
36	20	720	25920
44	16	704	30976
52	9	468	24336
	$\Sigma f = 60$	$\Sigma fx = 2272$	$\Sigma fx^2 = 91072$

$\bar{x} = 2272/60 = 37.9$, $s = 9.24$.

(b) From cumulative frequency polygon, median = 38, inter-quartile range = 45 − 32 − 13.

(c) Measure of skewness = 3(37.9 − 38)/9.24 = −0.03, not great, so prefer $\bar{x}$ and s.

7. (a)

x	f	fx	fx^2
0	21	0	0
1	13	13	13
2	8	16	32
3	3	9	27
4	4	16	64
5	1	5	25
	$\Sigma f = 50$	$\Sigma fx = 59$	$\Sigma fx^2 = 161$

$\bar{x} = 1.18$, $s = 1.37$.

(b) From this grouped frequency distribution table, median = 1, inter-quartile range = 2 − 0 = 2.
(c) Measure of skewness = 3(1.18 − 1)/1.37 = 0.39, not great, so prefer $\bar{x}$ and s.

8. (a)

x	f	fx	fx^2
2.5	60	150	375.00
7.5	100	750	5 625.00
12.5	75	937.5	11 718.75
17.5	45	787.5	13 781.25
22.5	15	337.5	7 593.75
27.5	4	110.0	3 025.00
32.5	1	32.5	1 056.25
	$\Sigma f = 300$	$\Sigma fx = 3105$	$\Sigma fx^2 = 43\,175.00$

$\bar{x} = 10.35$, $s = 6.08$.

(b) From cumulative frequency polygon, median = 9.5, inter-quartile range = 14.3 − 5.7 = 8.6.
(c) Measure of skewness = 3(10.35 − 9.5)/6.08 = 0.42, not great, so prefer $\bar{x}$ and s.

9. (a)

x	f	fx	fx^2
2	1	2	4
3	7	21	63
4	6	24	96
5	8	40	200
6	8	48	288
7	2	14	98
8	2	16	128
9	3	27	243
10	0	0	0
11	2	22	242
12	2	24	288
13	0	0	0
14	0	0	0
15	0	0	0
16	1	16	156
	$\Sigma f = 42$	$\Sigma fx = 254$	$\Sigma fx^2 = 1906$

$\bar{x} = 6.0$, $s = 3.0$.
(b) From this grouped frequency distribution table, median = 5, inter-quartile range = 7.25 − 4 = 3.25.

(c) Measure of skewness = 3(6 − 5)/3 = 1, quite large, so prefer median and inter-quartile range.

10.

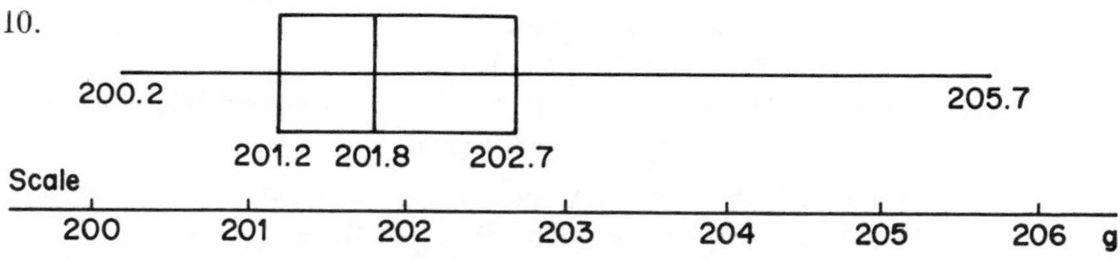

WORKSHEET 5 (Solutions)

1. Refer to Section 5.3 and Section 5.4.

2. Using $P(E') = 1 - P(E)$, where E is success, E' is 'not success', i.e. failure, P(failure) = 1 − 0.2 = 0.8.

3. The 3 events; 2 heads, 1 head and 1 tail, 2 tails, are not equally likely. P(HH) = P(HT) = P(TH) = P(TT) = 1/2 × 1/2 = 1/4.
   So P(2 heads) = P(HH) = 1/4, P(1 head) = P(HT or TH) = P(HT) + P(TH) = 1/2, P(2 tails) = P(TT) = 1/4.

4. Our estimate of the probability of heads is 1 after 5 tosses, but the number of trials is not large enough for an accurate estimate.

5. (a) Each die has same probability $(= \frac{1}{3})$ of being selected.
   (b) Y1, Y2, . . . , Y6, B1, B2, . . . , B6, G1, G2, . . . , G6; 18 outcomes in all.
   (c) Yes.
   (d) 1/18.
   (e) (i) 6/18, (ii) 3/18, (iii) 9/18, (iv) 3/18, (v) 12/18.

6. 13/50 = Area of rectangle with base 175 − 185/Total area of histogram
   (a) 13/50 × 100 = 26%, (b) 100%.

7. Suppose the number of U's after, 1, 2, 3, 4, 5, 10, 20, 30, 40, and 50 tosses is as follows:

No. of tosses	1	2	3	4	5	10	20	30	40	50
No. of U's	1	1	2	3	3	5	11	17	23	29
Est. of P(U)	1	0.5	0.67	0.75	0.60	0.50	0.55	0.57	0.58	0.58

(see Fig. B.6)

Fig. B.6

8. (a) A′, (b) A|B, (c) B|A.

9. (a) Probability of event A given that event B has occurred.
   (b) Probability of event B given that event A has occurred.
   (c) Probability that event A will not occur.
   (d) The probability of event A is not affected by whether event B has occurred.
   (e) If event A occurs, B cannot, and vice versa.

10. P(A and B) = P(A)P(B|A), or P(A and B) = P(B)P(A|B).
    If A and B are statistically independent, P(A and B) = P(A)P(B).

11. P(A or B or both) = P(A) + P(B) − P(A and B).
    If A and B are mutually exclusive, P(A or B) = P(A) + P(B).

12. (a) A and B are statistically independent.
    (b) A and B are mutually exclusive.

13. P(3 or 6) = P(3) + P(6) = 1/6 + 1/6 = 1/3.

14. P(red or picture or both) = P(red) + P(picture) − P(red and picture)
    = 26/52 + 16/52 − 8/52
    = 34/52.

15. The probability tree has 8 branch-ends, each having probability 1/8.
    (a) 1/8, (b) 3/8, (c) 3/8, (d) 1/8.

16. P(at least one hit) = 1 − P(all miss) = 1 − 1/2 × 2/3 × 3/4 = 3/4.

17. P(exactly one defective) = P(DD′D′D′ or D′DD′D′ or D′D′DD′ or D′D′D′D)
    = P(DD′D′D′) + P(D′DD′D′) + P(D′D′DD′) + P(D′D′D′D)
    = 0.03 × 0.97 × 0.97 × 0.97 plus three similar terms
    = 0.1095.

18. From information given, using C = car, H = house, use laws of probability with P(H) = 0.4, P(C) = 0.7, P(H and C) = 0.3.
    (a) P(H or C or both) = P(H) + P(C) − P(H and C) = 0.4 + 0.7 − 0.3 = 0.8.
    (b) Since P(H and C) = P(C) × P(H|C), 0.3 = 0.7 × P(H|C). Therefore P(H|C) = 0.3/0.7 = 0.43.

19.

	Bedroom type *S*	*D*	
Bath	2	9	11
No bath	4	5	9
	6	14	20

(a) 11/20 = 0.55.
(b) 2/11 = 0.18.

20. First member seems to be using P(U_1 and U_2) = P(U_1)P(U_2) = 1/4 × 1/4 where U_1, U_2 = unoccupied on first visit, second visit respectively.

Therefore P(occupied on either 1st or 2nd visit) $= 1 - (\frac{1}{4})^2 = 15/16$.
Incorrect because independence assumed, and it is reasonable to assume that if a house is unoccupied on one visit, the chances of it being unoccupied on second visit will be affected.
Second member seems to be adding the 1/4's. Is he using $P(U_1 \text{ or } U_2) = P(U_1) + P(U_2)$? Incorrect because U_1 and U_2 are not mutually exclusive.
Correct argument $P(U_1 \text{ and } U_2) = P(U_1)P(U_2|U_1) = 1/4P(U_2|U_1)$.
Therefore required probability $= 1 - 1/4P(U_2|U_1)$.
We don't know $P(U_2|U_1)$ but it must be between 0 (if unoccupied houses on 1st visit are all different from those unoccupied on 2nd visit) and 1 (if unoccupied houses on 1st visit are all exactly the same as those unoccupied on 2nd visit). Therefore required probability lies between $1 - 1/4 \times 0$ and $1 - 1/4 \times 1$, i.e. between 1 and 3/4.

21. (a) P(success) = P(lift-off success) × P(separation success|lift-off success) × P(mission completed|separation success)
$= (1 - 0.1)(1 - 0.05)(1 - 0.03)$
$= 0.8294$.
(b) P(failure) $= 1 - 0.8294 = 0.1706$.

WORKSHEET 6 (Solutions)

1. n is the number of trials, p is the probability of success in a single trial.
2. By checking the four conditions, see Section 6.3.
3. The choice is arbitrary, although to use Table D.1 it is better to choose the outcome with a probability of less than 0.5 as the 'success'.
4. The number of successes in n trials.
5. (a) 2, (b) 3, (c) 4, (d) 1.5, 0.866 (e) yes.
6. 0.0625, 0.2500, 0.3750, 0.2500, 0.0625. Because the five events are mutually exclusive and exhaustive. Expect 12 (or 13), 50, 75, 50, 13 (or 12), families.
7. $B(20, 0.2)$, (a) 4, (b) 0.0321, (c) 0.9885.
8. (a) 0.9231, (b) 0.2611, (c) 0.0002.
9. 0.0312, 0.1563, 0.3125, 0.3125, 0.1562, 0.0313 (using Table D.1).
10. 0.3828, probability.
11. (a) Because the four conditions are satisfied.
(b) $n = 20, p = 0.2$.
(c) (i) 0.0115, (ii) 0.9885.
12. $B(6, 0.1)$, (a) 0.5314, (b) 0.4686, where trial is 'examining an egg'. A 'success' is 'cracked egg'.
$B(5, 0.5314)$ where a trial is now 'examining a box', and 'success' is 'a box with no cracked eggs'. P(3 or more) = P(3) + P(4) + P(5) = 0.5587.
13. m is mean number of random events per unit time or space.
e is a constant, the number 2.718...

x is the number of random events per unit time or space.
x can take values, 0, 1, 2,..., ∞.

14. Standard deviation = 2, variance = 4. Mean and variance are equal in value.
15. In each unit of time (or space) the probability that the event will occur is the same.
16. $m = 1/10 = 0.1$ breakdowns per week. 0.9048, 0.0905, 0.0045.
17. $m = 1/5 = 0.2$ misprints per page. P(0) = 0.8187 or 81.87%. Expect no misprints on 409 (= 0.8187 × 500) pages.
18. $m = 2.5$, (a) 8.21%, (b) 45.62%, (c) P(4 or more) = 0.2424 or 24.24%.
19. $m = 3$, (a) 0.0498, (b) 0.2240, (c) 0.5768.
20. 0.634, 0.6321. Good agreement, first is correct, second is only an approximation, but an excellent one.
21. $m = np = 4$, P(0) = 0.0183.
22. (a) 0.0245, 0.0318, not such a good agreement. Answers using Poisson approximation are less accurate for small probabilities.
    (b) 0.1849, 0.1755, good agreement.

WORKSHEET 7 (Solutions)

1. Mean, standard deviation.
2. No, always symmetrical.
3. The total area under the curve for the normal distribution is 1, the total area of the rectangle for the rectangular distribution is 1.
4. (a) 4.75%, (b) 0.05%, (c) 95.2%. Expect 2, 0, 48 (to nearest whole orange).
   With $\mu = 70 - 5 = 65$, $\sigma = 3$, new answers are: (a) 0.05%, (b) 4.75%, (c) 95.2%. Expect 0, 2, 48.
5. For the 7 grades, percentages are 0.62%, 6.06%, 24.17%, 38.30%, 24.17%, 6.06%, 0.62%. Total price of 10 000 oranges = (62 × 4 + 606 × 5 + ··· + 62 × 10) = 70 000 p. Mean price = 7 p.

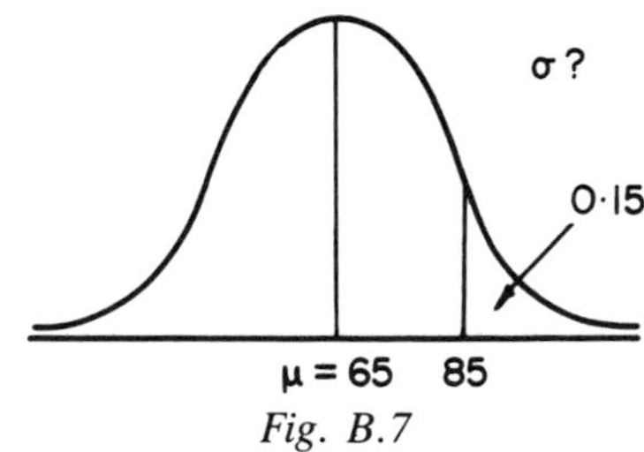

Fig. B.7

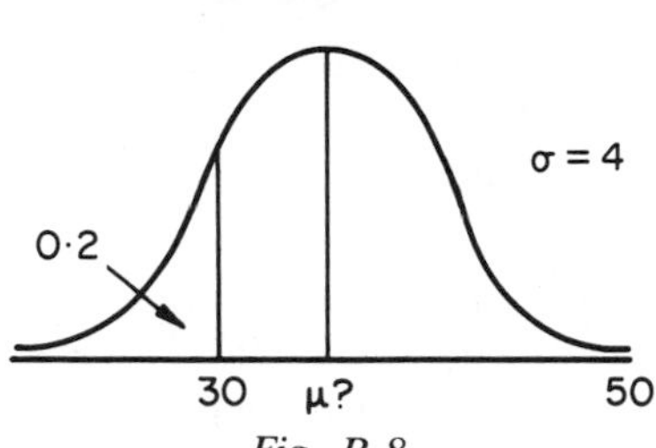

Fig. B.8

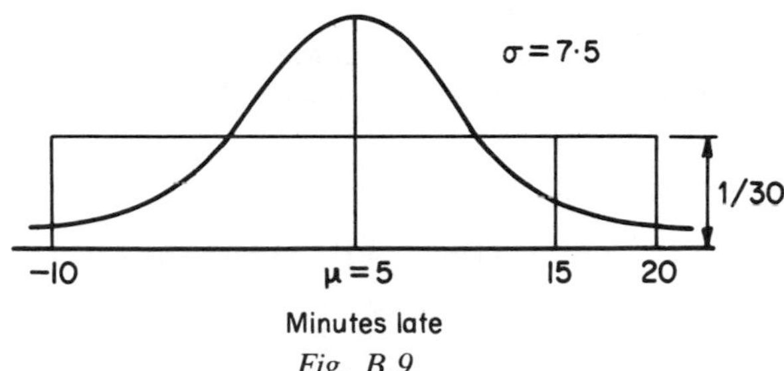

Fig. B.9

6. Percentage rejected = 18.15%. With new mean of 0.395, percentage rejected = 13.36%. This is a minimum since 0.395 is halfway between the rejection values of 0.38 and 0.41.

7. (a) 203, (b) 19, (c) 778.

8. For left-handed area of $1 - 0.15 = 0.85$, $z = 1.04$ using Table D.3(a) in reverse therefore $(x - \mu)/\sigma = (85 - 65)/\sigma = 1.04$, $\sigma = (85 - 65)/1.04 = 19.2$ cm. See Fig. B.7.
For 50 cm, $z = -0.78$, so 21.8% of years have less than 50 cm, using Table D.3.

9. For left-handed area of 0.2, $z = -0.84$. Therefore $(x - \mu)/\sigma = (30 - \mu)/4 = -0.84$. $\mu = 30 + 4 \times 0.84 =$ £33.36. See Fig. B.8.
For £50, $z = 4.16$ and from Table D.3(a) area to left of $z = 4.16$ is virtually 1. So no staff earn more than £50 a week.

10. (a) 50%, (b) 95.25%, (c) 99.95%.

11. $z = -2.33$, $(x - \mu)/\sigma = (x - 172)/8 = -2.33$, $x = 172 - 8 \times 2.33 = 153.4$ cm.

12. (a) 50%, (b) 95.45%. New target mean is 25.82 kg, 0.1% exceed 27.37 kg ($z = 3.1$).

13. (a) $z = (x - \mu)/\sigma = [(\mu + \sigma) - \mu]/\sigma = 1$, and $z = [(\mu - \sigma) - \mu]/\sigma = -1$. Hence area $= 0.8413 - (1 - 0.8413) = 0.6826$.
(b) Similar to (a) with $z = \pm 2$.
(c) Similar to (a) with $z = \pm 3$.
(d) Similar to (a) with $z = \pm 1.96$.
(e) Similar to (a) with $z = +1.645$.

14. Use normal approximation to the binomial, $n = 60, p = 0.8$, so $\mu = 48, \sigma = 3.1$.
    (a) P(>49.5) = 1 − 0.6844 = 0.3156.
    (b) P(49.5 − 50.5) = 0.1066.
    (c) P(<49.5) = 0.6844.

15. With rectangular distribution, area to right of 15 = 1/6. Late 1 day in 6. With normal distribution, $z = 1.33$, area to left of 15 = 0.9082, and hence area to right of 15 = 0.0918. Late 1 day in 11 (approximately). See Fig. B.9.

WORKSHEET 8 (Solutions)

1. (a), (b), (c), (d) – see Sections 8.1, 8.2 and 8.3.
   (e) A census is a 100% sample, so that the whole population is included in the sample, often the main purpose is to count the total number of individuals in the population.

2. See Section 8.3.

3. (a) Might catch slowest and largest first. Better to number mice 1 to 20 and use random numbers.
   (b) (i) Travellers more affluent than average adult.
   (ii) Shoppers for food more representative of adults, might be biased in favour of housewives, etc.
   (iii) Adults leaving job centre may be unemployed, might be biased in connection with unemployment.
   (c) We do not know how investigator actually 'randomly threw a quadrat'. Better to use grid method (see Fig. 8.1), and place quadrat with its centre at points chosen.
   (d) Use stratified sampling since there are three strata. Select 4, 5, 1 at random from the three types of hotel, respectively.
   (e) Initial sample correctly chosen, but many of those chosen may not be on the telephone. Even if there is a telephone in the house, only one name appears in the telephone directory. Party worker should visit those selected (even though there are problems – see Worksheet 5, Question 20).
   (f) People who visit doctor's surgery are not typical of a random sample of his possible patients. Also those who volunteer may do so because they are prone to influenza, again not random. Better to choose a random sample from the alphabetic list of patients, assign half randomly to the vaccine and the other half to a placebo.

4. Theory indicates the following discrete probability distribution for samples of size 2:

Mean of 2 dice scores	1	1.5	2.0	2.5	3.0	3.5	4.0	4.5	5.0	5.5	6.0
Probability	$\frac{1}{36}$	$\frac{2}{36}$	$\frac{3}{36}$	$\frac{4}{36}$	$\frac{5}{36}$	$\frac{6}{36}$	$\frac{5}{36}$	$\frac{4}{36}$	$\frac{3}{36}$	$\frac{2}{36}$	$\frac{1}{36}$
Expected frequency in 108 throws	3	6	9	12	15	18	15	12	9	6	3

Plot actual frequency against mean to show triangular shape.

WORKSHEET 9 (Solutions)

1. To give some measure of precision to a single value (or point) estimate of a population parameter such as the mean.
2. $\bar{x}$, s, n, and Table D.5. If n small, variable must be approximately normal.
3. False.
4. True.
5. True.
6. It does not imply '95% probability' (since either the population mean actually does lie between 10 and 12 and then the probability is 1 or it does not and then the probability is 0). Since, in repeated sampling, 95% of the 95% confidence intervals we calculate actually contain the mean, we feel that the confidence interval '10 to 12' has a very good chance of being one of those intervals which actually contains the mean. Think of betting on a horse at odds of 19 to 1 on. We can think of 'taking a risk of 5%' that the interval '10 to 12' does not contain the mean.
7. £28 to £32. Need to sample 400 customers.
8. 72% to 88%. Need to ask 6150 people.
9. 125.3 to 132.5, (a) 7.2, (b) 9.5, (c) 3.6.
10. 9220 pebbles.
11. $\bar{x} = 308.3$, $s = 131.9$, 274.2 to 342.4.
12. 0.67 to 6.75 kg. The assumption that difference in weights is approximately normal is reasonable since weight is approximately normal. Also a cross-diagram of the data indicates approximate normality (symmetry plus concentration of points in the middle).
13. $s = 0.0982$, -0.37 to -0.58 for $(\mu_A - \mu_B)$. Assumptions: (i) Percentages are normally distributed, (ii) $\sigma_A = \sigma_B$ (reasonable here since s_A and s_B are similar) (also see solutions to Questions 16, 17, 18 of Worksheet 10).
14. $\bar{d} = -10.45$, $s_d = 10.13$, -3.6 to -17.3 for μ_d, where d = 'Town A − Town B' rainfall. Assumption that d is normal, which looks reasonable from a cross-diagram.
15. $s = 4.00$, -4.0 to -14.3 for $(\mu_A - \mu_B)$. Assumptions: (i) Weights are normally distributed, reasonable if reasons for small variations are numerous and independent (see Section 7.2). (ii) $\sigma_A = \sigma_B$. (Refer to solution to Question 13 above.)
    (a) 440.7 to 448.6.
    (b) 449.4 to 458.3.
    Reasonable for B, not for A, since 95% confidence interval for B does contain 452, but 95% confidence interval for A does not contain 452. (See also Section 10.16, where this use of confidence intervals is discussed.)

WORKSHEET 10 (Solutions)

1. See Sections 10.2 and 10.3.

2, 3, 4. See Sections 10.4 and 10.14.

5. See Section 10.3.

6. If we wish to decide whether the value of the parameter is greater than (or less than) a particular value, then the alternative hypothesis is one-sided. If we wish to decide whether the value of the parameter is different from (i.e. not equal to) a particular value, so the direction of the difference is not of interest, then the alternative hypothesis is two-sided.

7. *Calc* $t = 4.37$, *Tab* $t = 2.228$, data do not support stated hypothesis.
Assumption Weight of sugar is 'approximately' normal. Reasonable if reasons for small variations in sugar weights are numerous and independent (see Section 7.2), or we could draw a cross-diagram (see Fig. 3.9) to see if we have the symmetry and concentration in middle of a normal distribution.

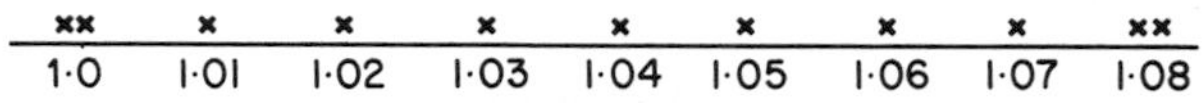

××	×	×	×	×	×	×	×	××
1·0	1·01	1·02	1·03	1·04	1·05	1·06	1·07	1·08

Fig. B.10

Figure B.10 shows reasonably even spread, and hence is not severely non-normal.

8. *Calc* $t = 10.54$, *Tab* $t = 1.645$, manufacturer's claim not justified.
Assumption Approximate normality, but not important here because of large sample size.

9. $\bar{x} = 87.67$, $s = 10.81$, *Calc* $t = -1.18$, *Tab* $t = 1.699$, data do not support claim.
Assumption Approximate normality of wages, which a histogram would indicate. In any case sample size is quite large.

10. *Calc* $z = 1.09$, *Tab* $z = 1.645$. Market share has not increased significantly.
Assumption The four binomial assumptions, main one being independence which means here that individuals do not influence each other in their choice of brand.

11. *Calc* $z = 3.46$, *Tab* $z = 1.96$. It is not reasonable to expect that 50% of all gourmets prefer thin soup.
Assumption Similar to Question 10.

12. *Calc* $z = -2.55$, *Tab* $z = 1.645$. It is reasonable to assume a lower death rate than 14%.
Assumption The four binomial conditions, main one being independence which means that chance of death is not influenced by others dying – probably true except for epidemics of fatal diseases (the Great Plague for example).

13. (a) Growing conditions within a farm will be more homogeneous than between farms.

(b) $\bar{d} = 0.3143$, $s_d = 0.4140$, *Calc* $t = 2.01$, *Tab* $t = 2.447$. Mean yields not significantly different.
Assumption Approximately normal differences; reasonable if reasons for small variations in differences are numerous and independent (see Section 7.2). As in Question 7, a cross-diagram could be drawn, but would not be conclusive.

14. $\bar{d} = -0.66$, $s_d = 0.8591$, *Calc* $t = -1.72$, *Tab* $t = 2.132$. Allegation not supported by data.
Assumption Approximately normal differences. No information on why variations in differences occur, and very little data. t-test dodgy here.

15. $\bar{d} = 0.72$, $s_d = 0.7941$, *Calc* $t = 2.87$, *Tab* $t = 1.833$. Drug gives significantly more hours sleep. Similar cross-diagram to Fig. B.10.

16. $\bar{x}_1 = 4673.3$, $s_1 = 120.94$, $\bar{x}_2 = 4370.0$, $s_2 = 214.48$, $s = 174.1$, *Calc* $t = 3.02$, *Tab* $t = 2.228$. There is a significant difference in mean strengths.
Assumptions (a) normality (approx.), difficult to tell here and very little data. (b) $\sigma_1 = \sigma_2$. s_1 and s_2 do not agree that well. Really need to do an F-test by comparing *Calc* $F = s_2^2/s_1^2 = 3.15$ (to give a number > 1) with *Tab* F which is 5.05 for samples of 6 and 6 respectively. Since *Calc* $F <$ *Tab* F, do not reject $H_0: \sigma_1 = \sigma_2$. Unfortunately this F test is very dependent on normality assumption (F test not covered in this book nor are F tables given. Refer, for example, to Chatfield (1983)).

17. $s = 0.5523$, *Calc* $t = 7.55$, *Tab* $t = 1.68$. It is reasonable to suppose corner shops are charging more on average than supermarkets.
Assumptions (a) Approximate normality, not important here because of large sample sizes. (b) $\sigma_1 = \sigma_2$. As in Question 16, F test would show this is a reasonable assumption.

18. $\bar{x}_A = 91.54$, $s_A = 23.27$, $\bar{x}_B = 104.77$, $s_B = 24.05$, $s = 23.66$, *Calc* $t = -3.19$, *Tab* $t = 1.645$. Mean amount of vanadium for area A significantly less than for area B.
Assumptions (a) Approximate normality, not important here since large sample sizes. (b) $\sigma_A = \sigma_B$, s_A and s_B are in very close agreement so this can be assumed to be reasonable.

19. (7) 1.02 to 1.06. Reject $H_0: \mu = 1$ since 1 outside 95% confidence interval.
(11) 54% to 66%. Reject $H_0: p = 0.5$, since 50% is outside confidence interval.
(13) -0.07 to 0.70. Do not reject $H_0: \mu_d = 0$, since 0 is inside the 95% confidence interval.
(16) 79.3 to 527.3. Reject $H_0: \mu_1 = \mu_2$, since 0 is outside the 95% confidence interval.

WORKSHEET 11 (Solutions)

1. Hypotheses, assumptions.
2. Assumptions, powerful.
3. Null, alternative, higher.

4. (a) Preference testing example,
   (b) Example where magnitudes of differences are known. Wilcoxon preferred.
5. Unpaired samples, powerful, assumptions, assumptions, normally, standard deviations.
6. Sign test for median, *calc. prob.* = $(\frac{1}{2})^9 = 0.002 < 0.025$. Median significantly different from 1 kg. Agrees with *t*-test conclusion of mean significantly different from 1 kg (mean and median are equal for normal distribution).
7. The number of cigarettes with nicotine content > 0.30 milligrams.
8. *Calc* $T = 3$, *Tab* $T = 0$. Median of amount of dye washed out for old dye not significantly less than for new dye.
9. (a) Wilcoxon, *Calc* $T = 1\frac{1}{2}$, *Tab* $T = 2$. Difference is significant.
   (b) Mann–Whitney, *U*, *Calc* $U = 26\frac{1}{2}$, *Tab* $U = 13$. Difference is not significant.
10. Mann–Whitney, *Calc* $U = 46$, *Tab* $U = 23$. Compounds equally effective.
11. Mann–Whitney, *Calc* $U = 9$, *Tab* $U = 37$. Significant difference between brands.

WORKSHEET 12 (Solutions)

1. Numerical.
2. Contingency, individuals, observed.
3. Independent.
4. Expected, E = row total × column total/grand total.
5. 5,

$$\frac{\Sigma(O - E)^2}{E},$$

   except for 2 × 2 table when we calculate

$$\frac{\Sigma(|O - E| - \frac{1}{2})^2}{E},$$

   rejection.
6. $(r - 1)(c - 1)$, 1.
7. 3.84.
8. *Calc* $\chi^2 = 2.91$, *Tab* $\chi^2 = 3.84$. Proportion of burnt out tubes for rented sets is not significantly different from proportion for bought sets.
9. *Calc* $\chi^2 = 10.34$, *Tab* $\chi^2 = 7.82$. Proportion of male to female is significantly different for the four garages. Tendency for females to prefer garage B.

10. Combine rows '30% – 70%' and 'under 30%' to form a 2 × 2 table (low E values). *Calc* $\chi^2 = 1.52$, *Tab* $\chi^2 = 3.84$. The chance of passing is independent of attendance (as defined by >70% or ≤70%).

11. *Calc* $\chi^2 = 5.48$, *Tab* $\chi^2 = 5.99$. Proportions of A, B, C not significantly different.

12. *Calc* $\chi^2 = 0.69$, *Tab* $\chi^2 = 3.84$, reasonable to assume independence of method of feeding and occurrence of the disease.

13. (a) 3 × 3 table, *Calc* $\chi^2 = 80.7$, *Tab* $\chi^2 = 9.49$. Opinions of men, women and children are significantly different. Largest effect is that more children like the flavour than expected under independence, but other large effects.
    (b) 3 × 2 table, *Calc* $\chi^2 = 5.8$, *Tab* $\chi^2 = 5.99$. Opinions of men and women are not significantly different.

WORKSHEET 13 (Solutions)

1. Scatter diagram.
2. Correlation coefficient.
3. Correlation coefficient, Pearson, r, ϱ.
4. r, arbitrary.
5. −1 to +1, negative, zero.
6. ϱ, normally distributed, uncorrelated.
7. If x = % increase in unemployment, y = % increase in manufacturing output,

$$\Sigma x = 70,\ \Sigma x^2 = 1136,\ \Sigma y = -50,\ \Sigma y^2 = 664,\ \Sigma xy = -732$$

   $n = 10$, $r = -0.739$, *Calc* $t = -3.10$, *Tab* $t = 1.86$. There is significant negative correlation, assuming normality of x and y which may be reasonable for these data, see Section 7.2. Difficult to draw other conclusions, except that both effects may be due to third and other factors such as world slump. In each country there will be individual factors as well, such as lack of investment, government policy, welfare state, attitudes to work and so on.

8. Number of times commercial shown is not normal. Use Spearman's $r_s = 0.744$ for these data. *Tab* $r_s = 0.643$, showing significant positive correlation. Increase in number of times commercial shown associated with increase in receipts. Scatter diagram gives impression that the effect is flattening off after about 30 commercials in the month.

9. No evidence of non-normality or normality. Safer with Spearman's $r_s = 0.473$, *Tab* $r_s = 0.714$. Larger areas not significantly associated with longer river lengths.

10. No evidence of non-normality or normality. $r_s = -0.964$, *Tab* $r_s = 0.714$. Lower death rate is significantly associated with higher percentage using filtered

water. Lower death rate could be due to other factors such as public awareness of need to boil unfiltered water, and better treatment of typhoid, etc.

11. Cross-diagrams for income and savings indicate that, while the distribution of income is reasonably normal, the distribution of savings may not be because of positive skewness. Safer to use Spearman's r_s here. $r_s = -0.08$. *Tab* $r_s = 0.377$, indicating no significant correlation between income and savings.
    There are two outliers in the scatter diagram but the effect of leaving them out (which we should not do without a good reason) would probably make no difference to the conclusion above.

WORKSHEET 14 (Solutions)

1. Predict, y, x.
2. Scatter, regression, straight line.
3. Did you beat 4000?
4. $\Sigma x = 229$, $\Sigma x^2 = 5333$, $\Sigma y = 3700$, $\Sigma y^2 = 1\,390\,200$, $\Sigma xy = 85\,790$, $n = 10$, $b = 11.92$, $a = 97.0$. Predict (a) 335, (b) 395, (c) 455. Last is the least reliable being furthest from the mean value of x, 22.9. Also 30 is just outside range of x values in data.
5. $\Sigma x = 15.05$, $\Sigma x^2 = 55.0377$, $\Sigma y = 15.2$, $\Sigma y^2 = 55.04$, $\Sigma xy = 55.01$, $n = 6$, $b = 0.9766$, $a = 0.0836$. At zero, predict true depth of 0.08, $s_r = 0.1068$, 95% confidence interval for true depth at zero is $-$ 0.13 to 0.30, but negative depths are impossible, so quote 0 to 0.30.
6. $\Sigma x = 289$, $\Sigma x^2 = 12\,629$, $\Sigma y = 73$, $\Sigma y^2 = 839$, $\Sigma xy = 3234$, $n = 7$, $b = 0.316$, $a = -2.60$. At (a) $x = 0$, cannot use equation, (b) $x = 50$, 13.2%, (c) $x = 100$, cannot use equation. $s_r = 1.28$, *Calc* $t = 6.51$, *Tab* $t = 2.57$, slope is significantly different from zero.
7. $\Sigma x = 965$, $\Sigma x^2 = 78\,975$, $\Sigma y = 370$, $\Sigma y^2 = 11\,598$, $\Sigma xy = 30\,220$, $n = 12$, $b = 0.339$, $a = 3.55$. (a) at 65, 23.5 to 27.7, (b) at 80, 29.5 to 31.9, (c) at 95, 33.8 to 37.8, using $s_r = 1.78$.
8. $\Sigma x = 55$, $\Sigma x^2 = 385$, $\Sigma y = 740$, $\Sigma y^2 = 57{,}150$, $\Sigma xy = 3655$, $n = 10$, $b = -5.03$, $a = 101.7$. $s_r = 6.15$, *Calc* $t = -7.43$, *Tab* $t = 1.86$, slope is significantly less than zero. 95% confidence interval for β is -6.59 to -3.47.
9. $\Sigma x = 280$, $\Sigma x^2 = 14\,000$, $\Sigma y = 580$, $\Sigma y^2 = 43\,180.44$, $\Sigma xy = 22\,477$ $n = 8$, $b = 0.518$, $a = 54.4$. (a) 67.3, (b) 82.9, (c) 98.4, but this is extrapolation.
10. Scatter diagram shows no clear pattern, might expect negative correlation. In fact $r = -0.066$, which is not significant (*Calc* $t = -0.19$).
11. Scatter diagram y against $1/x$ looks linear. Let $z = 1/x$, then (to 2 dps) $\Sigma z = 8.71$, $\Sigma z^2 = 14.53$, $\Sigma y = 44.5$, $\Sigma y^2 = 313.55$, $\Sigma zy = 44.81$, $n = 7$. For $y = a + bz$, $b = -\ 2.86$, $a = 9.92$. For $x = 1$, $z = 1$, predict $y = 7.06$, 95% confidence interval is 6.76 to 7.37 ($s_r = 0.30$).
    Pearson's r for y and $1/x$ should be larger in magnitude but negative.

WORKSHEET 15 (Solutions)

1. *Calc* $\chi^2 = 0.47$, *Tab* $\chi^2 = 7.82$. Data consistent with theory.
2. (a) *Calc* $\chi^2 = 12.7$, *Tab* $\chi^2 = 11.1$, reject uniform distribution.
   (b) *Calc* $\chi^2 = 60.4$, *Tab* $\chi^2 = 11.1$, reject 2 : 1 : 5 : 4 : 5 : 3 distribution. Allowing for volume, significantly more accidents occur during hours of darkness than expected.
3. *Calc* $\chi^2 = 13.7$, *Tab* $\chi^2 = 5.99$. Reject independence, more farms on flat land, fewer on mountainous land than expected.
4. *Calc* $\chi^2 = 5.5$, *Tab* $\chi^2 = 5.99$. Data consistent with 1 : 2 : 1 hypothesis.
5. (a) *Calc* $\chi^2 = 9.5$, *Tab* $\chi^2 = 7.82$. Data not consistent with $B(3, 0.5)$ distribution.
   (b) *Calc* $\chi^2 = 3.8$, *Tab* $\chi^2 = 5.99$. Data consistent with $B(3, 0.4633)$ distribution, indicating significantly fewer boys than girls.
6. See Section 8.3 for grid method of selecting random points at which to place quadrats. Number the pebbles within a quadrat, and choose 10 using random number tables.
   Calc $\chi^2 = 1.9$, *Tab* $\chi^2 = 12.6$. Data consistent with $B(10, 0.4)$ distribution.
7. *Calc* $\chi^2 = 5.5$, *Tab* $\chi^2 = 11.1$. Data consistent with $B(20, 0.1)$ distribution.
8. See Section 8.3 for grid method of selecting 80 random points. Estimated m is 1.8, *Calc* $\chi^2 = 16.6$, *Tab* $\chi^2 = 7.82$. Data not consistent with random distribution.
9. *Calc* $\chi^2 = 3.9$, *Tab* $\chi^2 = 9.49$. Data consistent with a random distribution ($m = 2$).
10. (a) $\bar{x} = 2$, $s^2 = 4.13$, so not approximately equal, so not reasonable to assume Poisson.
    (b) *Calc* $\chi^2 = 95.3$, *Tab* $\chi^2 = 7.82$. Data not consistent with random distribution. Many more cars than expected with either no defects or at least 4 defects.
11. $\bar{x} = 20$, $s = 5.74$, *Calc* $\chi^2 = 0.65$, *Tab* $\chi^2 = 3.84$. Data consistent with normal.
12. $\bar{x} = 201.96$, $s = 1.09$, *Calc* $\chi^2 = 1.4$, *Tab* $\chi^2 = 3.84$. Data consistent with normal.
13. $\bar{x} = 56.7$, $s = 22.0$, *Calc* $\chi^2 = 38.5$, *Tab* $\chi^2 = 12.6$. Data inconsistent with normal. Histogram suggests negative skewness, or possibly two distributions, one with mean around score of 20, another with mean around 65.

WORKSHEET 16 (Solutions)

1. (a) **HISTOGRAM C1**;
   INCREMENT 1;
   START 0.

Mean = 1.18, standard deviation = 1.366, median = 1, inter-quartile range = 2 − 0 = 2.

(b) Mean = 6.048, standard deviation = 3.004, median = 5, inter-quartile range = 7.25 − 4 = 3.25.

(c) Mean = 201.98, standard deviation = 1.06, median = 201.76, inter-quartile range = 202.72 − 201.26 = 1.46.

For histogram shapes and frequencies (counts) see solutions to Worksheet 3, Question 2.

2. **PDF**; with subcommands

BINOMIAL	**N = 4, P = 0.5.**	(Question 6)
BINOMIAL	**N = 5, P = 0.5.**	(Question 9)
POISSON	**MEAN = 0.1.**	(Question 16)
POISSON	**MEAN = 0.2.**	(Question 17)

3. **CDF**; with subcommands

BINOMIAL N = 50, P = 0.05. (Question 8)

(a) P(≥1) = 1 − P(0) = 1 − P(≤0) = 1 − 0.0769 = 0.9231
(b) P(2) = P(≤2) − P(≤1) = 0.5405 − 0.2794 = 0.2611
(c) P(≥10) = 1 − P(≤9) = 1 − 0.9998 = 0.0002

POISSON MEAN = 3. (Question 19)

(a) P(0) = P(≤0) = 0.0498
(b) P(3) = P(≤3) − P(≤2) = 0.6472 − 0.4232 = 0.2240
(c) P(≤3) = 1 − P(≤2) = 1 − 0.4232 = 0.5768

4. **RANDOM 100 C1**; with subcommand **BINOMIAL N = 20, P = 0.2**. Then use **DESCRIBE C1**. Expect mean (np) to be 4 and standard deviation ($\sqrt{np(1-p)}$ to be 1.79.

5. **CDF 75**; with subcommand **NORMAL MU = 70, SIGMA = 3**, gives area to left of 75 as 0.9522. So 0.0478 (4.78%) is to the right of 75. Similarly for **CDF 60**; **CDF 15**; with subcommand **NORMAL MU = 15, SIGMA = 3**, and so on.

6. **INVCDF 0.01**; with subcommand **NORMAL MU = 172, SIGMA = 8**. gives 153.4 cm.

7. **RANDOM 50 C1**; with subcommand **NORMAL MU = 70, SIGMA = 3**.

8. For samples of size 2, type in program given and then add:

```
MTB > RANDOM 1000 C4 C5;
SUBC > DISCRETE C1 C2.
MTB > RMEAN C4 C5 INTO C6
MTB > PRINT C6
MTB > DESCRIBE C6
MTB > HISTOGRAM C6
```

Mean and standard deviation should be close to 3.5 and $1.71/\sqrt{2} = 1.21$. Similar program for samples of size 3, mean and standard deviation should be close to 3.5 and $1.71/\sqrt{3} = 0.99$. Histogram shape should become more normal as sample size increases.

9. 0.67 to 6.75, following the example in Table 16.11. −0.370 to −0.580 for

$\mu_A - \mu_B$, using the **TWOSAMPLE-T** command (see Section 16.12 and Table 16.15).

10. *Calc* $t = -1.72$, $P = 0.08$, following example in Table 16.14.

11. *Calc* $\chi^2 = 4.00$, without Yates's continuity correction. Compare *Calc* $\chi^2 = 2.91$ in solution to Worksheet 12, Question 8.
    Calc $\chi^2 = 10.34$
    For the 3×2 table, 2 cells with expected frequencies (counts) less than 5. Combining rows 2 and 3, Minitab gives *Calc* $\chi^2 = 2.38$, without Yates's continuity correction. Compare *Calc* $\chi^2 = 1.52$ in solution to Worksheet 12, Question 10.

12. Use **READ** or **SET** to put the data into columns 1 and 2, then **PLOT C1 C2**. **CORRELATION C1 C2** gives $r = -0.739$.

13. Read the 'weight' data into C1, the 'consumption' data into C2, followed by
    MTB > BRIEF 1
    MTB > PLOT C2 C1
    MTB > REGRESSION C2 1 C1;
    SUBC > PREDICT 65;
    SUBC > PREDICT 80;
    SUBC > PREDICT 95.
    Minitab program for Worksheet 14, Question 9 is very similar to the above.

14. Expected frequencies are 312.75, 104.25, 104.25 and 34.75. *Calc* $\chi^2 = 0.47$, using same commands as in Table 16.20.
    Expected frequencies are 1.21, 8.06, 24.19, 43.00, 50.16, 40.13, 22.30, 8.49, 2.13, 0.31, 0.02, using same commands as in Table 16.21. However, several expected frequencies are less than 5 and it is necessary to combine bottom two categories and also top four categories. Then writing a program in which these expected and observed frequencies are read into C1 and C2 followed by
    MTB > LET C3 = (C2 − C1)2/C1**
    MTB > LET K1 = SUM (C3)
    MTB > PRINT K1
    gives K1 = *Calc* $\chi^2 = 1.89$.

MULTIPLE CHOICE TEST (Solutions)

1.c	2.b	3.c	4.b	5.b
6.b	7.a	8.b	9.c	10.b
11.b	12.c	13.a	14.a	15.c
16.b	17.c	18.b	19.a	20.a
21.a	22.b	23.a	24.a	25.b
26.c	27.a	28.c	29.a	30.c
31.b	32.b	33.c	34.b	35.c
36.a	37.c	38.a	39.c	40.a
41.b	42.c	43.a	44.a	45.c
46.c	47.b	48.c	49.b	50.a

Appendix C

GLOSSARY OF SYMBOLS

The references in brackets refer to the main chapter and section in which the symbol is introduced, defined or used.

ROMAN SYMBOLS

a	Intercept of sample regression line (14.1)
b	Slope of sample regression line (14.1)
$B(n, p)$	General binomial distribution (6.3)
c	Number of columns in contingency table (12.2)
d	Difference between pairs of values (9.10)
d	Difference between pairs of ranks (13.5)
$\bar{d}$	Sample mean difference (9.10, 10.12)
e	2.718 (2.4, 6.9)
E	Event (5.3)
E	Expected frequency (12.3, 15.2)
E'	Not E, complement of event E (5.12)
f	Frequency (4.2)
H_0	Null hypothesis (10.1)
H_1	Alternative hypothesis (10.1)
m	Mean number of random events per unit time or space (6.9)
n	Number of values in sample (= sample size) (2.1, 9.1, 9.6, 9.9)
n	Total number of equally likely outcomes (5.3)
n	(Large) number of trials (5.3)
n	Number of trials in a binomial experiment (6.3)
n	Number of pairs (9.10, 10.12)
n	Number of individuals in correlation and regression (13.2, 14.2)
$n!$	Factorial n (2.2)
$\binom{n}{x}$	$\dfrac{n!}{x!(n-x)!}$ (6.3)
$N(\mu, \sigma^2)$	General normal distribution (7.4)
O	Observed frequency (12.3, 15.2)

p	Probability of success in a single trial of a binomial experiment (6.3)
$P(E)$	Probability that event E will occur (5.3)
$P(E_1 \text{ and } E_2)$	Probability that both events E_1 and E_2 will occur (5.8)
$P(E_1 \text{ or } E_2 \text{ or both})$	Probability that either or both of events E_1 and E_2 will occur (5.8)
$P(E_2 \mid E_1)$	Probability that event E_2 will occur, given E_1 has occurred (5.9)
$P(x)$	Probability of x successes in binomial experiment (6.3)
$P(x)$	Probability of x random events per unit time or space (6.9)
r	Number of trials resulting in event E (5.3)
r	Number of equally likely outcomes resulting in event E (5.3)
r	Number of rows in contingency table (12.2)
r	Sample value of Pearson's correlation coefficient (13.2)
r_s	Sample value of Spearman's correlation coefficient (13.5)
R_1, R_2	Sum of ranks of values in samples of sizes n_1, n_2 in Mann–Whitney U test (11.7)
s	Sample standard deviation (4.7)
s^2	Sample variance (4.10)
s^2	Pooled estimate of variance (4.10)
s_d	Sample standard deviation of differences (9.10, 10.12)
s_r	Residual standard deviation (14.5)
t	Student's statistic (9.4, 9.10, 10.9)
T	Wilcoxon signed rank statistic (11.5)
T_+, T_-	Sum of ranks of positive and negative differences in Wilcoxon signed rank test (11.5)
U	Mann–Whitney statistic (11.7)
U_1, U_2	Values calculated in Mann–Whitney U test (11.7)
x	Any variable (2.1, 6.3, 6.9, 13.2)
x	Variable used to predict y variable in regression (14.1)
x_1, x_2, x_n, x_i	The first, second, nth, ith, values of x (2.1)
x_0	A particular value of x in regression (14.5)
$\bar{x}$	Sample mean of x (2.1, 4.2)
y	Variable used with variable x in correlation (13.2)
y	Variable to be predicted from x variable in regression (14.1)
$\bar{y}$	Sample mean of y (14.1)
z	Value used in Table D.3 (areas of normal distribution) (7.3)

GREEK SYMBOLS

α (alpha)	Probability used in t tables (9.4)
α	Probability used in χ^2 tables (12.3)
α	Intercept of population regression line (14.6)
β (beta)	Slope of population regression line (14.6)
μ (mu)	Mean of a normal distribution (7.2)
μ	Population mean (8.5, 9.2, 10.3)
μ_d	Mean of a population of differences (9.10, 10.12)
$\mu_{\bar{x}}$	Mean of distribution of $\bar{x}$ (8.5)

ν (nu)	Degrees of freedom (9.4, 9.7, 10.9, 10.13, 12.3, 13.3)
ϱ (rho)	Population value of Pearson's correlation coefficient (13.2)
σ (sigma)	Standard deviation of a normal distribution (7.2)
σ	Population standard deviation (8.5)
$\sigma_{\bar{x}}$	Standard deviation of distribution of $\bar{x}$ (8.5)
Σ (sigma)	Operation of summing (2.1)
χ^2 (chi-squared)	Chi-squared statistic (12.1, 15.1)

Appendix D

STATISTICAL TABLES

Table D.1 Cumulative binomial probabilities
Table D.2 Cumulative Poisson probabilities
Table D.3(a) Normal distribution function
Table D.3(b) Upper percentage points for the normal distribution
Table D.4 Random numbers
Table D.5 Percentage points of the t-distribution
Table D.6 Values of T for the Wilcoxon signed rank test
Table D.7 Values of U for the Mann–Whitney U test
Table D.8 Percentage points of the χ^2 distribution
Table D.9 Values of Spearman's r_s

ACKNOWLEDGEMENTS

The author would like to thank the following authors and publishers for their kind permission to adapt from the following tables:

Tables D.3 and D.5 from:

Pearson, E.S. and Hartley, H.O. (1966), *Biometrika Tables for Statisticians*, Vol. 1, 3rd edition, Cambridge University Press, Cambridge.

Table D.6 from:

Runyon, R.P. and Haber, A. (1968), *Fundamentals of Behavioural Statistics*, Addison-Wesley, Reading, Mass.; based on values in Wilcoxon, F., Katti, S. and Wilcox, R.A. (1963), *Critical Values and Probability Levels for the Wilcoxon Rank Sum Test and the Wilcoxon Signed Rank Test*, American Cyanamid Co., New York; and Wilcoxon, F. and Wilcox, R.A. (1964), *Some Rapid Approximate Statistical Procedures*, American Cyanamid Co., New York.

Table D.7 from:

Owen, D.B. (1962), *Handbook of Statistical Tables*, Addison-Wesley, Reading, Mass.; based on values in Auble, D. (1953), Extended tables for the Mann–Whitney Statistic. *Bulletin of the Institute of Educational Research at Indiana University*, **1**, 2.

Table D.8 from:

Mead, R. and Curnow, R.N. (1983), *Statistical Methods in Agriculture and Experimental Biology*, Chapman and Hall, London.

Table D.9 from:
Runyon, R.P. and Haber, A. (1968), *Fundamentals of Behavioural Statistics*, Addison-Wesley, Reading, Mass.; based on values in Olds, E.G. (1949), The 5% significance levels of sums of squares of rank differences and a correction. *Annals of Mathematical Statistics*, **20**, 117–18, and Olds, E.G. (1938), Distribution of the sum of squares of rank differences for small numbers of individuals. *Annals of Mathematical Statistics*, **9**, 133–48.

TABLE D.1 CUMULATIVE BINOMIAL PROBABILITIES

The table gives the probability of obtaining r or fewer successes in n independent trials, where p = probability of successes in a single trial.

	p =	0.01	0.02	0.03	0.04	0.05	0.06	0.07	0.08	0.09
$n = 2$	$r = 0$	.9801	.9604	.9409	.9216	.9025	.8836	.8649	.8464	.8281
	1	.9999	.9996	.9991	.9984	.9975	.9964	.9951	.9936	.9919
	2	1.0000	1.0000	1.0000	1.0000	1.0000	1.0000	1.0000	1.0000	1.0000
$n = 5$	$r = 0$	.9510	.9039	.8587	.8154	.7738	.7339	.6957	.6591	.6240
	1	.9990	.9962	.9915	.9852	.9774	.9681	.9575	.9456	.9326
	2	1.0000	.9999	.9997	.9994	.9988	.9980	.9969	.9955	.9937
	3		1.0000	1.0000	1.0000	1.0000	.9999	.9999	.9998	.9997
	4						1.0000	1.0000	1.0000	1.0000
$n = 10$	$r = 0$	.9044	.8171	.7374	.6648	.5987	.5386	.4840	.4344	.3894
	1	.9957	.9838	.9655	.9418	.9139	.8824	.8483	.8121	.7746
	2	.9999	.9991	.9972	.9938	.9885	.9812	.9717	.9599	.9460
	3	1.0000	1.0000	.9999	.9996	.9990	.9980	.9964	.9942	.9912
	4			1.0000	1.0000	.9999	.9998	.9997	.9994	.9990
	5					1.0000	1.0000	1.0000	1.0000	.9999
	6									1.0000
$n = 20$	$r = 0$	.8179	.6676	.5438	.4420	.3585	.2901	.2342	.1887	.1516
	1	.9831	.9401	.8802	.8103	.7358	.6605	.5869	.5169	.4516
	2	.9990	.9929	.9790	.9561	.9245	.8850	.8390	.7879	.7334
	3	1.0000	.9994	.9973	.9926	.9841	.9710	.9529	.9294	.9007
	4		1.0000	.9997	.9990	.9974	.9944	.9893	.9817	.9710
	5			1.0000	.9999	.9997	.9991	.9981	.9962	.9932
	6				1.0000	1.0000	.9999	.9997	.9994	.9987
	7						1.0000	1.0000	.9999	.9998
	8								1.0000	1.0000
$n = 50$	$r = 0$	.6050	.3642	.2181	.1299	.0769	.0453	.0266	.0155	.0090
	1	.9106	.7358	.5553	.4005	.2794	.1900	.1265	.0827	.0532
	2	.9862	.9216	.8108	.6767	.5405	.4162	.3108	.2260	.1605
	3	.9984	.9822	.9372	.8609	.7604	.6473	.5327	.4253	.3303
	4	.9999	.9968	.9832	.9510	.8964	.8206	.7290	.6290	.5277
	5	1.0000	.9995	.9963	.9856	.9622	.9224	.8650	.7919	.7072
	6		.9999	.9993	.9964	.9882	.9711	.9417	.8981	.8404
	7		1.0000	.9999	.9992	.9968	.9906	.9780	.9562	.9232

Continued

p =	0.01	0.02	0.03	0.04	0.05	0.06	0.07	0.08	0.09
8			1.0000	.9999	.9992	.9973	.9927	.9833	.9672
9				1.0000	.9998	.9993	.9978	.9944	.9875
10					1.0000	.9998	.9994	.9983	.9957
11						1.0000	.9999	.9995	.9987
12							1.0000	.9999	.9996
13								1.0000	.9999
14									1.0000

Continued

	p =	0.10	0.15	0.20	0.25	0.30	0.35	0.40	0.45	0.50
$n = 2$	$r = 0$	.8100	.7225	.6400	.5625	.4900	.4225	.3600	.3025	.2500
	1	.9900	.9775	.9600	.9375	.9100	.8775	.8400	.7975	.7500
	2	1.0000	1.0000	1.0000	1.0000	1.0000	1.0000	1.0000	1.0000	1.0000
$n = 5$	$r = 0$	.5905	.4437	.3277	.2373	.1681	.1160	.0778	.0503	.0313
	1	.9185	.8352	.7373	.6328	.5282	.4284	.3370	.2562	.1875
	2	.9914	.9734	.9421	.8965	.8369	.7648	.6826	.5831	.5000
	3	.9995	.9978	.9933	.9844	.9692	.9460	.9130	.8688	.8125
	4	1.0000	.9999	.9997	.9990	.9976	.9947	.9898	.9815	.9688
	5		1.0000	1.0000	1.0000	1.0000	1.0000	1.0000	1.0000	1.0000
$n = 10$	$r = 0$	.3487	.1969	.1074	.0563	.0282	.0135	.0060	.0025	.0010
	1	.7361	.5443	.3758	.2440	.1493	.0860	.0464	.0233	.0107
	2	.9298	.8202	.6778	.5256	.3828	.2616	.1673	.0996	.0547
	3	.9872	.9500	.8791	.7759	.6496	.5138	.3823	.2660	.1719
	4	.9984	.9901	.9672	.9219	.8497	.7515	.6331	.5044	.3770
	5	.9999	.9986	.9936	.9803	.9527	.9051	.8338	.7384	.6230
	6	1.0000	.9999	.9991	.9965	.9894	.9740	.9452	.8980	.8281
	7		1.0000	.9999	.9996	.9984	.9952	.9877	.9726	.9453
	8			1.0000	1.0000	.9999	.9995	.9983	.9955	.9893
	9					1.0000	1.0000	.9999	.9997	.9990
	10							1.0000	1.0000	1.0000
$n = 20$	$r = 0$	.1216	.0388	.0115	.0032	.0008	.0002			
	1	.3917	.1756	.0692	.0243	.0076	.0021	.0005	.0001	
	2	.6769	.4049	.2061	.0913	.0355	.0121	.0036	.0009	.0002
	3	.8670	.6477	.4114	.2252	.1071	.0444	.0160	.0049	.0013
	4	.9568	.8298	.6296	.4148	.2375	.1182	.0510	.0189	.0059
	5	.9887	.9327	.8042	.6172	.4164	.2454	.1256	.0553	.0207
	6	.9976	.9781	.9133	.7858	.6080	.4166	.2500	.1299	.0577
	7	.9996	.9941	.9679	.8982	.7723	.6010	.4159	.2520	.1316
	8	.9999	.9987	.9900	.9591	.8867	.7624	.5956	.4143	.2517
	9	1.0000	.9998	.9974	.9861	.9520	.8782	.7553	.5914	.4119
	10		1.0000	.9994	.9961	.9829	.9468	.8725	.7507	.5881
	11			.9999	.9991	.9949	.9804	.9435	.8692	.7483
	12			1.0000	.9998	.9987	.9940	.9790	.9420	.8684
	13				1.0000	.9997	.9985	.9935	.9786	.9423
	14					1.0000	.9997	.9984	.9936	.9793
	15						1.0000	.9997	.9985	.9941
	16							1.0000	.9997	.9987
	17								1.0000	.9998
	18									1.0000

Continued

$p =$		0.10	0.15	0.20	0.25	0.30	0.35	0.40	0.45	0.50
$n = 50$	$r = 0$	.0052	.0003							
	1	.0338	.0029	.0002						
	2	.1117	.0142	.0013	.0001					
	3	.2503	.0460	.0057	.0005					
	4	.4312	.1121	.0185	.0021	.0002				
	5	.6161	.2194	.0480	.0070	.0007	.0001			
	6	.7702	.3613	.1034	.0194	.0025	.0002			
	7	.8779	.5188	.1904	.0453	.0073	.0008	.0001		
	8	.9421	.6681	.3073	.0916	.0183	.0025	.0002		
	9	.9755	.7911	.4437	.1637	.0402	.0067	.0008	.0001	
	10	.9906	.8801	.5836	.2622	.0789	.0160	.0022	.0002	
	11	.9968	.9372	.7107	.3816	.1390	.0342	.0057	.0006	
	12	.9990	.9699	.8139	.5110	.2229	.0661	.0133	.0018	.0002
	13	.9997	.9868	.8894	.6370	.3279	.1163	.0280	.0045	.0005
	14	.9999	.9947	.9393	.7481	.4468	.1878	.0540	.0104	.0013
	15	1.0000	.9981	.9692	.8369	.5692	.2801	.0955	.0220	.0033
	16		.9993	.9856	.9017	.6839	.3889	.1561	.0427	.0077
	17		.9998	.9937	.9449	.7822	.5060	.2369	.0765	.0164
	18		.9999	.9975	.9713	.8594	.6216	.3356	.1273	.0325
	19		1.0000	.9991	.9861	.9152	.7264	.4465	.1974	.0595
	20			.9997	.9937	.9522	.8139	.5610	.2862	.1013
	21			.9999	.9974	.9749	.8813	.6701	.3900	.1611
	22			1.0000	.9990	.9877	.9290	.7660	.5019	.2399
	23				.9996	.9944	.9604	.8438	.6134	.3359
	24				.9999	.9976	.9793	.9022	.7160	.4439
	25				1.0000	.9991	.9900	.9427	.8034	.5561
	26					.9997	.9955	.9686	.8721	.6641
	27					.9999	.9981	.9840	.9220	.7601
	28					1.0000	.9993	.9924	.9556	.8389
	29						.9997	.9966	.9765	.8987
	30						.9999	.9986	.9884	.9405
	31						1.0000	.9995	.9947	.9675
	32							.9998	.9978	.9836
	33							.9999	.9991	.9923
	34							1.0000	.9997	.9967
	35								.9999	.9987
	36								1.0000	.9995
	37									.9998
	38									1.0000

Continued

TABLE D.2 CUMULATIVE POISSON PROBABILITIES

The table gives the probability of r or fewer random events per unit time or space, when the average number of such events is m.

m =	0.1	0.2	0.3	0.4	0.5	0.6	0.7	0.8	0.9	1.0
r = 0	.9048	.8187	.7408	.6703	.6065	.5488	.4966	.4493	.4066	.3679
1	.9953	.9825	.9631	.9384	.9098	.8781	.8442	.8088	.7725	.7358
2	.9998	.9989	.9964	.9921	.9856	.9769	.9659	.9526	.9371	.9197
3	1.0000	.9999	.9997	.9992	.9982	.9966	.9942	.9909	.9865	.9810
4		1.0000	1.0000	.9999	.9998	.9996	.9992	.9986	.9977	.9963
5				1.0000	1.0000	1.0000	.9999	.9998	.9997	.9994
6							1.0000	1.0000	1.0000	.9999
7										1.0000

m =	1.1	1.2	1.3	1.4	1.5	1.6	1.7	1.8	1.9	2.0
r = 0	.3329	.3012	.2725	.2466	.2231	.2019	.1827	.1653	.1496	.1353
1	.6990	.6626	.6268	.5918	.5578	.5249	.4932	.4628	.4337	.4060
2	.9004	.8795	.8571	.8335	.8088	.7834	.7572	.7306	.7037	.6767
3	.9743	.9662	.9569	.9463	.9344	.9212	.9068	.8913	.8747	.8571
4	.9946	.9923	.9893	.9857	.9814	.9763	.9704	.9636	.9559	.9473
5	.9990	.9985	.9978	.9968	.9955	.9940	.9920	.9896	.9868	.9834
6	.9999	.9997	.9996	.9994	.9991	.9987	.9981	.9974	.9966	.9955
7	1.0000	1.0000	.9999	.9999	.9998	.9997	.9996	.9994	.9992	.9989
8			1.0000	1.0000	1.0000	1.0000	.9999	.9999	.9998	.9998
9							1.0000	1.0000	1.0000	1.0000

m =	2.1	2.2	2.3	2.4	2.5	2.6	2.7	2.8	2.9	3.0
r = 0	.1225	.1108	.1003	.0907	.0821	.0743	.0672	.0608	.0550	.0498
1	.3796	.3546	.3309	.3084	.2873	.2674	.2487	.2311	.2146	.1991
2	.6496	.6227	.5960	.5697	.5438	.5184	.4936	.4695	.4460	.4232
3	.8386	.8194	.7993	.7787	.7576	.7360	.7141	.6919	.6696	.6472
4	.9379	.9275	.9162	.9041	.8912	.8774	.8629	.8477	.8318	.8153
5	.9796	.9751	.9700	.9643	.9580	.9510	.9433	.9349	.9258	.9161
6	.9941	.9925	.9906	.9884	.9858	.9828	.9794	.9756	.9713	.9665
7	.9985	.9980	.9974	.9967	.9958	.9947	.9934	.9919	.9901	.9881
8	.9997	.9995	.9994	.9991	.9989	.9985	.9981	.9976	.9969	.9962
9	.9999	.9999	.9999	.9998	.9997	.9996	.9995	.9993	.9991	.9989
10	1.0000	1.0000	1.0000	1.0000	.9999	.9999	.9999	.9998	.9998	.9997
11					1.0000	1.0000	1.0000	1.0000	.9999	.9999
12									1.0000	1.0000

Continued

m =	3.1	3.2	3.3	3.4	3.5	3.6	3.7	3.8	3.9	4.0
r = 0	.0450	.0408	.0369	.0334	.0302	.0273	.0247	.0224	.0202	.0183
1	.1847	.1712	.1586	.1468	.1359	.1257	.1162	.1074	.0992	.0916
2	.4012	.3799	.3594	.3397	.3208	.3027	.2854	.2689	.2531	.2381
3	.6248	.6025	.5803	.5584	.5366	.5152	.4942	.4735	.4532	.4335
4	.7982	.7806	.7626	.7442	.7254	.7064	.6872	.6678	.6484	.6288
5	.9057	.8946	.8829	.8705	.8576	.8441	.8301	.8156	.8006	.7851
6	.9612	.9554	.9490	.9421	.9347	.9267	.9182	.9091	.8995	.8893
7	.9858	.9832	.9802	.9769	.9733	.9692	.9648	.9599	.9546	.9489

m =	3.1	3.2	3.3	3.4	3.5	3.6	3.7	3.8	3.9	4.0
8	.9953	.9943	.9931	.9917	.9901	.9883	.9863	.9840	.9815	.9786
9	.9986	.9982	.9978	.9973	.9967	.9960	.9952	.9942	.9931	.9919
10	.9996	.9995	.9994	.9992	.9990	.9987	.9984	.9981	.9977	.9972
11	.9999	.9999	.9998	.9998	.9997	.9996	.9995	.9994	.9993	.9991
12	1.0000	1.0000	1.0000	.9999	.9999	.9999	.9999	.9998	.9998	.9997
13				1.0000	1.0000	1.0000	1.0000	1.0000	.9999	.9999
14									1.0000	1.0000

m =	4.1	4.2	4.3	4.4	4.5	4.6	4.7	4.8	4.9	5.0
r = 0	.0166	.0150	.0136	.0123	.0111	.0101	.0091	.0082	.0074	.0067
1	.0845	.0780	.0719	.0663	.0611	.0563	.0518	.0477	.0439	.0404
2	.2238	.2102	.1974	.1851	.1736	.1626	.1523	.1425	.1333	.1247
3	.4142	.3954	.3772	.3594	.3423	.3257	.3097	.2942	.2793	.2650
4	.6093	.5898	.5704	.5512	.5321	.5132	.4946	.4763	.4582	.4405
5	.7693	.7531	.7367	.7199	.7029	.6858	.6684	.6510	.6335	.6160
6	.8786	.8675	.8558	.8436	.8311	.8180	.8046	.7908	.7767	.7622
7	.9427	.9361	.9290	.9214	.9134	.9049	.8960	.8867	.8769	.8666
8	.9755	.9721	.9683	.9642	.9597	.9549	.9497	.9442	.9382	.9319
9	.9905	.9889	.9871	.9851	.9829	.9805	.9778	.9749	.9717	.9682
10	.9966	.9959	.9952	.9943	.9933	.9922	.9910	.9896	.9880	.9863
11	.9989	.9986	.9983	.9980	.9976	.9971	.9966	.9960	.9953	.9945
12	.9997	.9996	.9995	.9993	.9992	.9990	.9988	.9986	.9983	.9980
13	.9999	.9999	.9998	.9998	.9997	.9997	.9996	.9995	.9994	.9993
14	1.0000	1.0000	1.0000	.9999	.9999	.9999	.9999	.9999	.9998	.9998
15				1.0000	1.0000	1.0000	1.0000	1.0000	.9999	.9999
16									1.0000	1.0000

Continued

m =	5.2	5.4	5.6	5.8	6.0	6.2	6.4	6.6	6.8	7.0
$r = 0$	.0055	.0045	.0037	.0030	.0025	.0020	.0017	.0014	.0011	.0009
1	.0342	.0289	.0244	.0206	.0174	.0146	.0123	.0103	.0087	.0073
2	.1088	.0948	.0824	.0715	.0620	.0536	.0463	.0400	.0344	.0296
3	.2381	.2133	.1906	.1700	.1512	.1342	.1189	.1052	.0928	.0818
4	.4061	.3733	.3422	.3127	.2851	.2592	.2351	.2127	.1920	.1730
5	.5809	.5461	.5119	.4783	.4457	.4141	.3837	.3547	.3270	.3007
6	.7324	.7017	.6703	.6384	.6063	.5742	.5423	.5108	.4799	.4497
7	.8449	.8217	.7970	.7710	.7440	.7160	.6873	.6581	.6285	.5987
8	.9181	.9027	.8857	.8672	.8472	.8259	.8033	.7796	.7548	.7291
9	.9603	.9512	.9409	.9292	.9161	.9016	.8858	.8686	.8502	.8305
10	.9823	.9775	.9718	.9651	.9574	.9486	.9386	.9274	.9151	.9015
11	.9927	.9904	.9875	.9841	.9799	.9750	.9693	.9627	.9552	.9467
12	.9972	.9962	.9949	.9932	.9912	.9887	.9857	.9821	.9779	.9730
13	.9990	.9986	.9980	.9973	.9964	.9952	.9937	.9920	.9898	.9872
14	.9997	.9995	.9993	.9990	.9986	.9981	.9974	.9966	.9956	.9943
15	.9999	.9998	.9998	.9996	.9995	.9993	.9990	.9986	.9982	.9976
16	1.0000	.9999	.9999	.9999	.9998	.9997	.9996	.9995	.9993	.9990
17		1.0000	1.0000	1.0000	.9999	.9999	.9999	.9998	.9997	.9996

m =	5.2	5.4	5.6	5.8	6.0	6.2	6.4	6.6	6.8	7.0
18					1.0000	1.0000	1.0000	.9999	.9999	.9999
19								1.0000	1.0000	1.0000

m =	7.2	7.4	7.6	7.8	8.0	8.2	8.4	8.6	8.8	9.0
$r = 0$	.0007	.0006	.0005	.0004	.0003	.0003	.0002	.0002	.0002	.0001
1	.0061	.0051	.0043	.0036	.0030	.0025	.0021	.0018	.0015	.0012
2	.0255	.0219	.0188	.0161	.0138	.0118	.0100	.0086	.0073	.0062
3	.0719	.0632	.0554	.0485	.0424	.0370	.0323	.0281	.0244	.0212
4	.1555	.1395	.1249	.1117	.0996	.0887	.0789	.0701	.0621	.0550
5	.2759	.2526	.2307	.2103	.1912	.1736	.1573	.1422	.1284	.1157
6	.4204	.3920	.3646	.3384	.3134	.2896	.2670	.2457	.2256	.2068
7	.5689	.5393	.5100	.4812	.4530	.4254	.3987	.3728	.3478	.3239
8	.7027	.6757	.6482	.6204	.5925	.5647	.5369	.5094	.4823	.4557
9	.8096	.7877	.7649	.7411	.7166	.6915	.6659	.6400	.6137	.5874
10	.8867	.8707	.8535	.8352	.8159	.7955	.7743	.7522	.7294	.7060
11	.9371	.9265	.9148	.9020	.8881	.8731	.8571	.8400	.8220	.8030
12	.9673	.9609	.9536	.9454	.9362	.9261	.9150	.9029	.8898	.8758
13	.9841	.9805	.9762	.9714	.9658	.9595	.9524	.9445	.9358	.9261
14	.9927	.9908	.9886	.9859	.9827	.9791	.9749	.9701	.9647	.9585
15	.9969	.9959	.9948	.9934	.9918	.9898	.9875	.9848	.9816	.9780
16	.9987	.9983	.9978	.9971	.9963	.9953	.9941	.9926	.9909	.9889
17	.9995	.9993	.9991	.9988	.9984	.9979	.9973	.9966	.9957	.9947
18	.9998	.9997	.9996	.9995	.9993	.9991	.9989	.9985	.9981	.9976
19	.9999	.9999	.9999	.9998	.9997	.9997	.9995	.9994	.9992	.9989
20	1.0000	1.0000	1.0000	.9999	.9999	.9999	.9998	.9998	.9997	.9996
21				1.0000	1.0000	1.0000	.9999	.9999	.9999	.9998
22							1.0000	1.0000	1.0000	.9999
23										1.0000

m =	9.2	9.4	9.6	9.8	10.0	11.0	12.0	13.0	14.0	15.0
r = 0	.0001	.0001	.0001	.0001						
1	.0010	.0009	.0007	.0006	.0005	.0002	.0001			
2	.0053	.0045	.0038	.0033	.0028	.0012	.0005	.0002	.0001	
3	.0184	.0160	.0138	.0120	.0103	.0049	.0023	.0011	.0005	.0002
4	.0486	.0429	.0378	.0333	.0293	.0151	.0076	.0037	.0018	.0009
5	.1041	.0935	.0838	.0750	.0671	.0375	.0203	.0107	.0055	.0028
6	.1892	.1727	.1574	.1433	.1301	.0786	.0458	.0259	.0142	.0076
7	.3010	.2792	.2584	.2388	.2202	.1432	.0895	.0540	.0316	.0180
8	.4296	.4042	.3796	.3558	.3328	.2320	.1550	.0998	.0621	.0374
9	.5611	.5349	.5089	.4832	.4579	.3405	.2424	.1658	.1094	.0699
10	.6820	.6576	.6329	.6080	.5830	.4599	.3472	.2517	.1757	.1185
11	.7832	.7626	.7412	.7193	.6968	.5793	.4616	.3532	.2600	.1848
12	.8607	.8448	.8279	.8101	.7916	.6887	.5760	.4631	.3585	.2676
13	.9156	.9042	.8919	.8786	.8645	.7813	.6815	.5730	.4644	.3632
14	.9517	.9441	.9357	.9265	.9165	.8540	.7720	.6751	.5704	.4657
15	.9738	.9691	.9638	.9579	.9513	.9074	.8444	.7636	.6694	.5681

m =	9.2	9.4	9.6	9.8	10.0	11.0	12.0	13.0	14.0	15.0
16	.9865	.9838	.9806	.9770	.9730	.9441	.8987	.8355	.7559	.6641
17	.9934	.9919	.9902	.9881	.9857	.9678	.9370	.8905	.8272	.7489
18	.9969	.9962	.9952	.9941	.9928	.9823	.9626	.9302	.8826	.8195
19	.9986	.9983	.9978	.9972	.9965	.9907	.9787	.9573	.9235	.8752
20	.9994	.9992	.9990	.9987	.9984	.9953	.9884	.9750	.9521	.9170
21	.9998	.9997	.9996	.9995	.9993	.9977	.9939	.9859	.9712	.9469
22	.9999	.9999	.9998	.9998	.9997	.9990	.9970	.9924	.9833	.9673
23	1.0000	1.0000	.9999	.9999	.9999	.9995	.9985	.9960	.9907	.9805
24			1.0000	1.0000	1.0000	.9998	.9993	.9980	.9950	.9888
25						.9999	.9997	.9990	.9974	.9938
26						1.0000	.9999	.9995	.9987	.9967
27							.9999	.9998	.9994	.9983
28							1.0000	.9999	.9997	.9991
29								1.0000	.9999	.9996
30									.9999	.9998
31									1.0000	.9999
32										1.0000

Continued

TABLE D.3(a) NORMAL DISTRIBUTION FUNCTION

For a normal distribution with a mean, μ, and standard deviation, σ, and a particular value of x, calculate $z = (x - \mu)/\sigma$. The table gives the area to the left of x, see Fig. D.1.

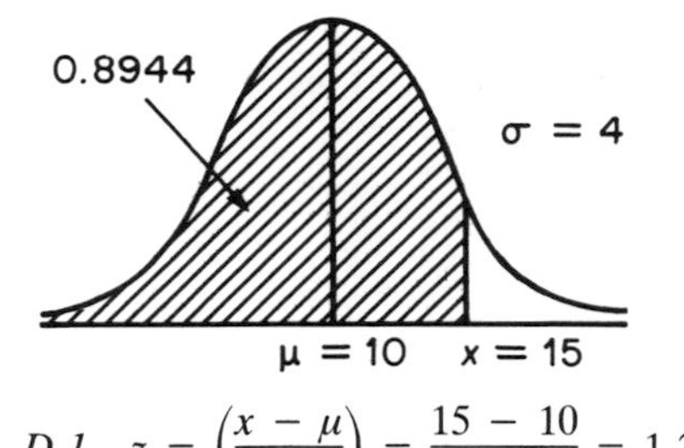

Fig. D.1 $z = \left(\frac{x - \mu}{\sigma}\right) = \frac{15 - 10}{4} = 1.25.$

$z = \frac{(x - \mu)}{\sigma}$	0.00	0.01	0.02	0.03	0.04	0.05	0.06	0.07	0.08	0.09
0.0	.5000	.5040	.5080	.5120	.5160	.5199	.5239	.5279	.5319	.5359
0.1	.5398	.5438	.5478	.5517	.5557	.5596	.5636	.5675	.5714	.5753
0.2	.5793	.5832	.5871	.5910	.5948	.5987	.6026	.6064	.6103	.6141
0.3	.6179	.6217	.6255	.6293	.6331	.6368	.6406	.6443	.6480	.6517
0.4	.6554	.6591	.6628	.6664	.6700	.6736	.6772	.6808	.6844	.6879
0.5	.6915	.6950	.6985	.7019	.7054	.7088	.7123	.7157	.7190	.7224
0.6	.7257	.7291	.7324	.7357	.7389	.7422	.7454	.7486	.7517	.7549
0.7	.7580	.7611	.7642	.7673	.7704	.7734	.7764	.7794	.7823	.7852
0.8	.7881	.7910	.7939	.7967	.7995	.8023	.8051	.8078	.8106	.8133
0.9	.8159	.8186	.8212	.8238	.8264	.8289	.8315	.8340	.8365	.8389
1.0	.8413	.8438	.8461	.8485	.8508	.8531	.8554	.8577	.8599	.8621
1.1	.8643	.8665	.8686	.8708	.8729	.8749	.8770	.8790	.8810	.8830
1.2	.8849	.8869	.8888	.8907	.8925	.8944	.8962	.8980	.8997	.9015
1.3	.9032	.9049	.9066	.9082	.9099	.9115	.9131	.9147	.9162	.9177
1.4	.9192	.9207	.9222	.9236	.9251	.9265	.9279	.9292	.9306	.9319
1.5	.9332	.9345	.9357	.9370	.9382	.9394	.9406	.9418	.9429	.9441
1.6	.9452	.9463	.9474	.9484	.9495	.9505	.9515	.9525	.9535	.9545
1.7	.9554	.9564	.9573	.9582	.9591	.9599	.9608	.9616	.9625	.9633
1.8	.9641	.9649	.9656	.9664	.9671	.9678	.9686	.9693	.9699	.9706
1.9	.9713	.9719	.9726	.9732	.9738	.9744	.9750	.9756	.9761	.9767

Continued

$z = \frac{(x - \mu)}{\sigma}$	0.00	0.01	0.02	0.03	0.04	0.05	0.06	0.07	0.08	0.09
2.0	.9772	.9778	.9783	.9788	.9793	.9798	.9803	.9808	.9812	.9817
2.1	.9821	.9826	.9830	.9834	.9838	.9842	.9846	.9850	.9854	.9857
2.2	.9861	.9864	.9868	.9871	.9875	.9878	.9881	.9884	.9887	.9890
2.3	.9893	.9896	.9898	.9901	.9904	.9906	.9909	.9911	.9913	.9916
2.4	.9918	.9920	.9922	.9925	.9927	.9929	.9931	.9932	.9934	.9936
2.5	.9938	.9940	.9941	.9943	.9945	.9946	.9948	.9949	.9951	.9952
2.6	.9953	.9955	.9956	.9957	.9959	.9960	.9961	.9962	.9963	.9964
2.7	.9965	.9966	.9967	.9968	.9969	.9970	.9971	.9972	.9973	.9974
2.8	.9974	.9975	.9976	.9977	.9977	.9978	.9979	.9979	.9980	.9981
2.9	.9981	.9982	.9982	.9983	.9984	.9984	.9985	.9985	.9986	.9986
3.0	.9987	.9987	.9987	.9988	.9988	.9989	.9989	.9989	.9990	.9990
3.1	.9990	.9991	.9991	.9991	.9992	.9992	.9992	.9992	.9993	.9993
3.2	.9993	.9993	.9994	.9994	.9994	.9994	.9994	.9995	.9995	.9995
3.3	.9995	.9995	.9995	.9996	.9996	.9996	.9996	.9996	.9996	.9997
3.4	.9997	.9997	.9997	.9997	.9997	.9997	.9997	.9997	.9997	.9998

TABLE D.3(b) UPPER PERCENTAGE POINTS FOR THE NORMAL DISTRIBUTION

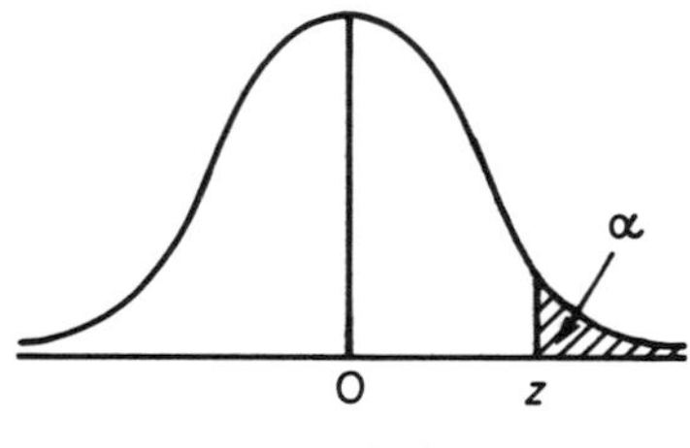

Fig. D.2

The table gives the values of z for various right hand tail areas, α, see Fig. D.2.

α	0.05	0.025	0.01	0.005	0.001	0.0005
z	1.645	1.96	2.33	2.58	3.09	3.29

TABLE D.4 RANDOM NUMBERS

30 89 34 43 98	38 51 15 30 26	02 57 93 32 67	19 91 72 23 06	59 24 11 06 50
79 50 49 98 07	05 88 29 05 29	73 15 65 17 92	26 05 21 60 73	55 48 97 54 50
53 64 54 20 36	05 26 90 12 98	73 98 56 47 60	44 54 45 97 21	25 70 96 58 72
87 23 75 21 50	54 47 46 35 72	11 66 30 44 63	69 50 82 74 58	98 25 68 47 79
91 54 58 41 48	70 11 94 79 12	36 63 12 52 72	43 41 11 52 98	91 77 91 85 00
92 41 24 08 42	64 96 82 07 01	40 00 95 09 30	23 40 08 19 78	55 50 92 84 96
65 63 25 34 62	93 01 96 23 23	81 31 94 09 02	75 98 27 85 59	53 09 94 37 37
93 64 13 39 70	98 38 71 77 89	47 98 47 22 09	98 85 91 86 42	30 60 34 07 23
92 44 97 54 10	53 06 50 66 76	13 89 09 41 28	93 04 75 68 09	78 22 82 88 10
69 37 57 14 85	43 72 12 89 80	07 01 17 91 30	17 00 49 53 99	46 51 26 74 28
88 13 45 79 30	32 44 38 84 94	26 65 83 04 43	88 70 99 09 89	31 59 08 29 11
30 86 16 00 13	89 22 16 01 29	98 65 92 13 36	26 88 58 18 89	67 19 71 92 28
19 39 94 95 22	70 99 77 50 29	30 16 69 87 18	48 56 34 92 85	42 54 25 72 84
04 01 90 59 21	33 16 80 53 51	90 02 92 76 72	03 82 77 75 72	33 44 87 58 29
17 45 23 69 94	53 68 59 13 13	68 39 80 62 31	70 44 32 01 47	54 43 70 97 08
13 35 10 58 52	66 73 38 05 80	45 71 76 21 80	10 58 72 17 06	50 72 97 41 48
07 48 12 02 82	51 55 21 61 13	44 27 63 97 04	56 13 88 48 02	34 15 84 30 87
08 16 12 72 05	72 10 63 76 44	92 84 98 81 43	71 66 24 27 16	06 32 39 21 89
51 94 42 32 70	21 82 38 94 46	59 34 75 61 97	72 76 50 50 30	70 27 08 16 72
06 78 72 46 93	36 77 57 19 49	99 18 26 11 63	74 29 96 14 57	76 72 92 86 28
39 14 12 52 96	24 33 70 06 77	56 59 42 11 80	33 05 63 40 14	22 70 62 17 05
71 31 34 36 97	98 57 79 44 68	06 62 74 23 69	77 41 05 17 26	41 68 37 19 53
57 64 15 98 66	13 41 98 06 19	64 53 36 19 16	19 90 71 70 74	04 03 30 05 34
64 26 20 69 40	12 85 65 75 73	92 57 43 97 70	71 28 02 89 91	86 98 64 56 73
91 38 37 54 09	99 35 01 78 03	09 53 57 79 53	50 23 00 90 49	45 28 45 00 94
89 29 45 54 07	22 17 50 32 64	07 30 41 19 36	32 18 08 94 48	20 84 02 47 95
81 31 03 44 27	43 93 91 10 38	72 95 27 58 65	02 23 61 23 17	17 70 26 19 79
05 45 30 21 51	05 14 61 37 61	47 39 50 22 73	28 06 14 72 89	53 64 75 09 70
03 61 43 09 65	35 22 77 22 50	50 37 79 34 14	65 03 56 93 62	34 03 93 18 14
82 75 76 86 14	93 52 73 37 68	83 46 04 11 96	24 14 84 07 19	88 54 05 04 29
62 91 08 18 91	52 65 53 89 39	95 43 21 88 25	36 97 60 89 07	12 03 57 31 39
99 61 53 27 31	18 30 38 21 32	91 03 04 61 53	19 81 45 69 05	35 63 25 00 53
44 29 75 03 84	52 19 73 07 26	92 21 25 48 18	98 14 24 72 12	26 24 89 86 53
51 17 94 61 54	16 39 17 30 32	41 23 37 62 20	51 62 33 79 66	51 95 89 43 55
87 51 27 95 72	31 82 22 31 18	20 31 03 93 60	50 93 18 75 26	62 64 57 46 85
58 12 50 48 30	85 34 65 89 19	63 58 41 42 56	03 67 41 69 48	81 13 44 42 70
78 25 85 91 28	01 85 26 47 58	66 11 84 77 18	30 47 19 42 74	80 13 53 72 66
97 09 87 30 35	04 26 88 10 58	18 44 75 06 52	92 49 73 70 79	49 42 20 09 96
69 08 45 81 37	89 68 51 99 15	33 07 14 39 61	78 05 50 34 14	72 32 78 30 59
82 74 69 78 50	51 47 00 57 40	51 84 26 51 23	14 08 30 96 92	56 71 54 59 96
71 08 26 53 23	43 60 71 41 63	95 26 14 78 09	73 74 63 73 21	06 79 69 81 90
17 60 07 10 21	77 42 60 77 01	20 14 04 09 89	55 79 97 62 57	13 59 38 42 41
90 07 13 82 73	77 37 58 21 35	29 81 98 80 85	51 58 49 82 66	46 94 59 42 25
14 04 16 79 09	72 01 15 51 47	01 12 32 87 84	65 27 89 34 07	40 57 95 06 77
42 44 93 98 30	13 10 61 85 30	46 82 99 79 93	48 62 46 26 71	19 98 34 48 28

TABLE D.5 PERCENTAGE POINTS OF THE *t*-DISTRIBUTION

For a t-distribution with ν degrees of freedom, the table gives the values of t which are exceeded with probability α. Figure D.3 shows a t-distribution with $\nu = 10$ df.

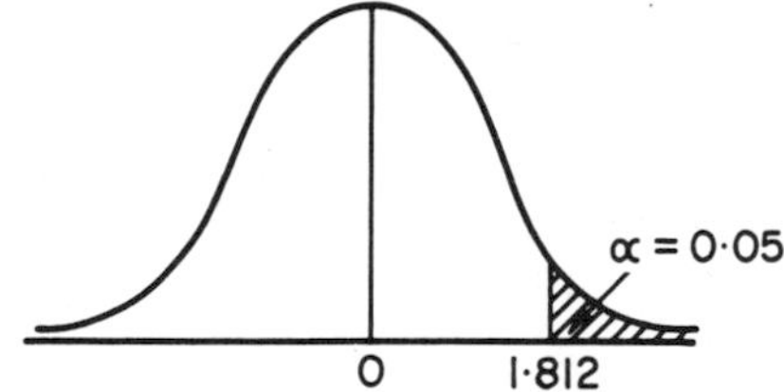

Fig. D.3 t-distribution with $\nu = 10$ df.

$\alpha =$	0.10	0.05	0.025	0.01	0.005	0.001	0.0005
$\nu = 1$	3.078	6.314	12.706	31.821	63.657	318.31	636.62
2	1.886	2.920	4.303	6.965	9.925	22.326	31.598
3	1.638	2.353	3.182	4.541	5.841	10.213	12.924
4	1.533	2.132	2.776	3.747	4.604	7.173	8.610
5	1.476	2.015	2.571	3.365	4.032	5.893	6.869
6	1.440	1.943	2.447	3.143	3.707	5.208	5.959
7	1.415	1.895	2.365	2.998	3.499	4.785	5.408
8	1.397	1.860	2.306	2.896	3.355	4.501	5.041
9	1.383	1.833	2.262	2.821	3.250	4.297	4.781
10	1.372	1.812	2.228	2.764	3.169	4.144	4.587
11	1.363	1.796	2.201	2.718	3.106	4.025	4.437
12	1.356	1.782	2.179	2.681	3.055	3.930	4.318
13	1.350	1.771	2.160	2.650	3.012	3.852	4.221
14	1.345	1.761	2.145	2.624	2.977	3.787	4.140
15	1.341	1.753	2.131	2.602	2.947	3.733	4.073
16	1.337	1.746	2.120	2.583	2.921	3.686	4.015
17	1.333	1.740	2.110	2.567	2.898	3.646	3.965
18	1.330	1.734	2.101	2.552	2.878	3.610	3.922
19	1.328	1.729	2.093	2.539	2.861	3.579	3.883
20	1.325	1.725	2.086	2.528	2.845	3.552	3.850
21	1.323	1.721	2.080	2.518	2.831	3.527	3.819
22	1.321	1.717	2.074	2.508	2.819	3.505	3.792
23	1.319	1.714	2.069	2.500	2.807	3.485	3.767
24	1.318	1.711	2.064	2.492	2.797	3.467	3.745
25	1.316	1.708	2.060	2.485	2.787	3.450	3.725
26	1.315	1.706	2.056	2.479	2.779	3.435	3.707

Continued

α =	0.10	0.05	0.025	0.01	0.005	0.001	0.0005
27	1.314	1.703	2.052	2.473	2.771	3.421	3.690
28	1.313	1.701	2.048	2.467	2.763	3.408	3.674
29	1.311	1.699	2.045	2.462	2.756	3.396	3.659
30	1.310	1.697	2.042	2.457	2.750	3.385	3.646
40	1.303	1.684	2.021	2.423	2.704	3.307	3.551
60	1.296	1.671	2.000	2.390	2.660	3.232	3.460
120	1.289	1.658	1.980	2.358	2.617	3.160	3.373
∞	1.282	1.645	1.960	2.326	2.576	3.090	3.291

TABLE D.6 VALUES OF *T* FOR THE WILCOXON SIGNED RANK TEST

	Level of significance for one-sided H_1			
	0.05	0.025	0.01	0.005
	Level of significance for two-sided H_1			
n	0.10	0.05	0.02	0.01
5	0	–	–	–
6	2	0	–	–
7	3	2	0	–
8	5	3	1	0
9	8	5	3	1
10	10	8	5	3
11	13	10	7	5
12	17	13	9	7
13	21	17	12	9
14	25	21	15	12
15	30	25	19	15
16	35	29	23	19
17	41	34	27	23
18	47	40	32	27
19	53	46	37	32
20	60	52	43	37
21	67	58	49	42
22	75	65	55	48
23	83	73	62	54
24	91	81	69	61
25	100	89	76	68

TABLE D.7 VALUES OF U FOR THE MANN–WHITNEY U TEST

Critical values of U for the Mann–Whitney test for 0.05 (first value) and 0.01 (second value) significance levels for two-sided H_1, and for 0.025 and 0.005 levels for one-sided H_1.

n_2 \ n_1	1	2	3	4	5	6	7	8	9	10	11	12	13	14	15	16	17	18	19	20
1	–	–	–	–	–	–	–	–	–	–	–	–	–	–	–	–	–	–	–	–
	–	–	–	–	–	–	–	–	–	–	–	–	–	–	–	–	–	–	–	–
2	–	–	–	–	–	–	–	0	0	0	0	1	1	1	1	1	2	2	2	2
	–	–	–	–	–	–	–	–	–	–	–	–	–	–	–	–	–	–	0	0
3	–	–	–	–	0	1	1	2	2	3	3	4	4	5	5	6	6	7	7	8
	–	–	–	–	–	–	–	–	0	0	0	1	1	1	2	2	2	2	3	3
4	–	–	–	0	1	2	3	4	4	5	6	7	8	9	10	11	11	12	13	14
	–	–	–	–	–	0	0	1	1	2	2	3	3	4	5	5	6	6	7	8
5	–	–	0	1	2	3	5	6	7	8	9	11	12	13	14	15	17	18	19	20
	–	–	–	–	0	1	1	2	3	4	5	6	7	7	8	9	10	11	12	13
6	–	–	1	2	3	5	6	8	10	11	13	14	16	17	19	21	22	24	25	27
	–	–	–	0	1	2	3	4	5	6	7	9	10	11	12	13	15	16	17	18
7	–	–	1	3	5	6	8	10	12	14	16	18	20	22	24	26	28	30	32	34
	–	–	–	0	1	3	4	6	7	9	10	12	13	15	16	18	19	21	22	24
8	–	0	2	4	6	8	10	13	15	17	19	22	24	26	29	31	34	36	38	41
	–	–	–	1	2	4	6	7	9	11	13	15	17	18	20	22	24	26	28	30
9	–	0	2	4	7	10	12	15	17	20	23	26	28	31	34	37	39	42	45	48
	–	–	0	1	3	5	7	9	11	13	16	18	20	22	24	27	29	31	33	36
10	–	0	3	5	8	11	14	17	20	23	26	29	33	36	39	42	45	48	52	55
	–	–	0	2	4	6	9	11	13	16	18	21	24	26	29	31	34	37	39	42

Continued

TABLE D.7 *contd.*

n_2 \ n_1	1	2	3	4	5	6	7	8	9	10	11	12	13	14	15	16	17	18	19	20
11	–	0	3	6	9	13	16	19	23	26	30	33	37	40	44	47	51	55	58	62
	–	–	0	2	5	7	10	13	16	18	21	24	27	30	33	36	39	42	45	48
12	–	1	4	7	11	14	18	22	26	29	33	37	41	45	49	53	57	61	65	69
	–	–	1	3	6	9	12	15	18	21	24	27	31	34	37	41	44	47	51	54
13	–	1	4	8	12	16	20	24	28	33	37	41	45	50	54	59	63	67	72	76
	–	–	1	3	7	10	13	17	20	24	27	31	34	38	42	45	49	53	57	60
14	–	1	5	9	13	17	22	26	31	36	40	45	50	55	59	64	69	74	78	83
	–	–	1	4	7	11	15	18	22	26	30	34	38	42	46	50	54	58	63	67
15	–	1	5	10	14	19	24	29	34	39	44	49	54	59	64	70	75	80	85	90
	–	–	2	5	8	12	16	20	24	29	33	37	42	46	51	55	60	64	69	73
16	–	1	6	11	15	21	26	31	37	42	47	53	59	64	70	75	81	86	92	98
	–	–	2	5	9	13	18	22	27	31	36	41	45	50	55	60	65	70	74	79
17	–	2	6	11	17	22	28	34	39	45	51	57	63	69	75	81	87	93	99	105
	–	–	2	6	10	15	19	24	29	34	39	44	49	54	60	65	70	75	81	86
18	–	2	7	12	18	24	30	36	42	48	55	61	67	74	80	86	93	99	106	112
	–	–	2	6	11	16	21	26	31	37	42	47	53	58	64	70	75	81	87	92
19	–	2	7	13	19	25	32	38	45	52	58	65	72	78	85	92	99	106	113	119
	–	0	3	7	12	17	22	28	33	39	45	51	57	63	69	74	81	87	93	99
20	–	2	8	14	20	27	34	41	48	55	62	69	76	83	90	98	105	112	119	127
	–	0	3	8	13	18	24	30	36	42	48	54	60	67	73	79	86	92	99	105

TABLE D.8 PERCENTAGE POINTS OF THE χ^2 DISTRIBUTION

For a χ^2 distribution with ν degrees of freedom, the table gives the values of χ^2 which are exceeded with probability α. See Fig. D.4.

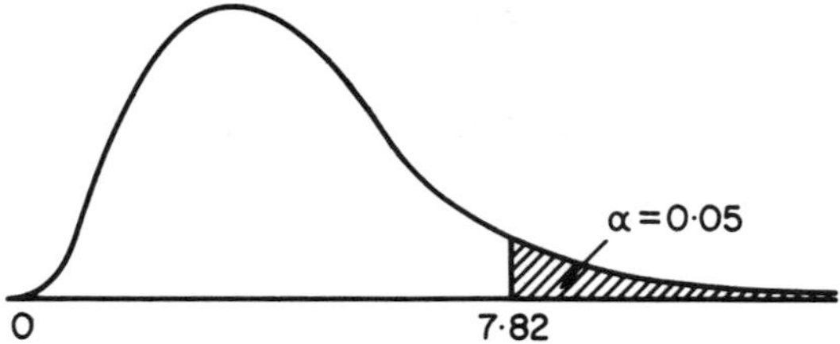

Fig. D.4 χ^2 distribution with ν = 3 df.

α =	0.50	0.10	0.05	0.025	0.01	0.001
ν = 1	0.45	2.71	3.84	5.02	6.64	10.8
2	1.39	4.61	5.99	7.38	9.21	13.8
3	2.37	6.25	7.82	9.35	11.3	16.3
4	3.36	7.78	9.49	11.1	13.3	18.5
5	4.35	9.24	11.1	12.8	15.1	20.5
6	5.35	10.6	12.6	14.5	16.8	22.5
7	6.35	12.0	14.1	16.0	18.5	24.3
8	7.34	13.4	15.5	17.5	20.1	26.1
9	8.34	14.7	16.9	19.0	21.7	27.9
10	9.34	16.0	18.3	20.5	23.2	29.6
12	11.3	18.5	21.0	23.3	26.2	32.9
15	14.3	22.3	25.0	27.5	30.6	37.7
20	19.3	28.4	31.4	34.2	37.6	45.3
24	23.3	33.2	36.4	39.4	43.0	51.2
30	29.3	40.3	43.8	47.0	50.9	59.7
40	39.3	51.8	55.8	59.3	63.7	73.4
60	59.3	74.4	79.1	83.3	88.4	99.6

TABLE D.9 VALUES OF SPEARMAN'S r_s

	Level of significance for one-sided H_1			
	0.05	0.025	0.01	0.005
	Level of significance for two-sided H_1			
n	0.1	0.05	0.02	0.01
5	0.900	1.000	1.000	–
6	0.829	0.886	0.943	1.000
7	0.714	0.786	0.893	0.929
8	0.643	0.738	0.833	0.881
9	0.600	0.683	0.783	0.833
10	0.564	0.648	0.746	0.794
12	0.506	0.591	0.712	0.777
14	0.456	0.544	0.645	0.715
16	0.425	0.506	0.601	0.665
18	0.399	0.475	0.564	0.625
20	0.377	0.450	0.534	0.591
22	0.359	0.428	0.508	0.562
24	0.343	0.409	0.485	0.537
26	0.329	0.392	0.465	0.515
28	0.317	0.377	0.448	0.496
30	0.306	0.364	0.432	0.478

Appendix E

FURTHER READING

The following books are mainly aimed at specific subject areas:

Allard, R.J. (1974) *An Approach to Econometrics*, Philip Allan, Oxford.

Armitage, P. (1971) *Statistical Methods in Medical Research*, Blackwell Scientific Publications, Oxford.

Bailey, N.T.J. (1969) *Statistical Methods in Biology*, Hodder, London.

Chatfield, C. (1983) *Statistics for Technology*, Chapman and Hall, London.

Croft, D. (1976) *Applied Statistics for Management Studies*, Macdonald and Evans, London.

Eckschlager, K. (1969) *Errors, Measurement and Results in Chemical Analysis*, Van Nostrand Reinhold, London.

Edwards, B. (1980) *Quantitative and Accounting Methods*, Macdonald and Evans, Plymouth.

Finney, D.J. (1980) *Statistics for Biologists*, Chapman and Hall, London.

Freund, J.E. and Williams, F.J. (1982) *Elementary Business Statistics*, Prentice-Hall, Englewood Cliffs, NJ.

Guilford, J.P. (1978) *Fundamental Statistics for Psychology and Education*, McGraw-Hill, New York.

Hammond, R. and McCullagh, P.S. (1978) *Quantitative Techniques in Geography*, Oxford University Press, Oxford.

Lark, P.D., Craven, B.R. and Bosworth, R.C.L. (1968) *The Handling of Chemical Data*, Pergamon, Oxford.

Mead, R. and Curnow, R.N. (1983) *Statistical Methods in Agriculture and Experimental Biology*, Chapman and Hall, London.

Norcliffe, G.B. (1977) *Inferential Statistics for Geographers*, Hutchinson Education, London.

Parker, R.E. (1979) *Introductory Statistics for Biology*, Edward Arnold, London.

Siegel, S. (1956) *Nonparametric Statistics (for the Behavioural Sciences)*, McGraw-Hill, New York.

Snodgrass, J.G. (1978) *The Numbers Game* (*Statistics for Psychology*), Oxford University Press, Oxford.

Sumner, J.R. (1972) *Statistics for Hotels and Catering*, Edward Arnold, London.

Till, R. (1974) *Statistical Methods for the Earth Scientist*, Macmillan, London.

Wetherill, G.B. (1981) *Intermediate Statistical Methods*, Chapman and Hall, London.

Index

A priori definition of probability 40
Addition law of probability 45
Alternative hypothesis 100, 101
  one-sided 101, 105
  two-sided 101
Approximation
  normal approximation to binomial 72
  Poisson approximation to binomial 60
Arithmetic mean, *see* Mean
Association
  of categorical variables, 130
  of numerical variables, *see* Correlation
Assumptions
  confidence interval estimation 96
  hypothesis testing 111, 126
Averages 24
  law of 40

Bar chart 18
Bernoulli trials 53
Bias in sampling 78
Binomial distribution 52
  goodness-of-fit test for 169
  mean of 55
  normal approximation to 72
  Poisson approximation to 60
  probabilities, calculation of 54
  probabilities, using Minitab 186
  standard deviation of 55
  table of probabilities 234
Box and whisker plots 34

Categorical variable 2, 130
Central limit theorem 82
Central tendency, measures of 24
Charts
  bar 18
  line 17
  pie 18
Chi-squared (χ^2) distribution
  table of percentage points, 247
Chi-squared (χ^2) test, *see* Hypothesis tests
Coefficient of variation 34
Complementary events 47
Confidence interval 85
  for binomial probability 91
  for difference in means of two populations 94
  for mean of a population of differences 93
  for means, using Minitab 190
  for population mean 86, 88
  for predicted value in regression 156
  for slope of regression line 165
  width of interval 87
Confidence level 85
Confidence limits 87
Contingency table 15, 130
Continuity correction
  in calculating z values 72, 120, 122, 125
  in 2 × 2 contingency table 132
Continuous
  probability distributions 65
  variable 2
Correlation 139
  connection with regression 160
  linear 143
  positive 142
  negative 142

non-linear 143
nonsense 145
rank 146
zero 142
Correlation coefficient 139
interpretation of 143
Pearson's product moment
calculation of 139
hypothesis test for 142
range of possible values 141, 148
using Minitab 195
Critical value 104, 106
Cross diagram 20
Cumulative frequency
polygon 16
table 13
Cumulative probabilities
binomial 56, 186
Poisson 60, 187

Data, statistical 2
Decimal places 8
Degrees of freedom 90
for χ^2 (Chi-squared) test for goodness-of-fit 169, 171, 173, 176
for χ^2 (Chi-squared) test of independence 133
for *t*
correlation coefficient 143
difference in means of two populations 95, 109
mean of population 88, 103
mean of population of differences 94, 108
predicted values in regression 157
slope of regression line 160
Diagrams
bar chart 18
box and whisker plots 34
cross 20
cumulative frequency polygon 16
histogram 16
line chart 17
Minitab 183, 195
ogive 16
pictogram 19
pie-chart 18
probability tree 47
scatter diagram 20
stem and leaf displays 20
time-series graph 20
tree diagram 47
Discrete
variable 2
probability distributions 52
Distribution
binomial 52
continuous probability 65
cumulative frequency 13
discrete probability 52
grouped frequency 11
normal 67
Poisson 58
rectangular (uniform) 71
sampling distribution of sample mean 81
simple proportion 167
t (Student's) 88, 89, 103, 142
Distribution-free tests 110

Equally likely outcomes 40
Error term 85, 88, 92
Estimation
interval 85
point 85
Events 40
complementary 47
exhaustive 46
mutually exclusive 46
random 57
statistically independent 44
Exhaustive events 46
Expected frequency 132, 168, 170, 172, 174
Experiment 40
Exploratory data analysis 20, 34
Exponential number, e 8

F test 223
Factorials 7
Fisher exact test 134
Frequency
cumulative 13
expected 132, 168, 170, 172, 174
grouped 11
observed 132, 168, 170, 172, 174

Goodness-of-fit test 167
  binomial 169, 197
  normal 173
  Poisson 171
  simple proportion 167, 197
Graphical methods 16, 183, 195
Grouped data
  inter-quartile range 33
  mean 25
  median 26
  mode 27
  standard deviation 31

Histogram 16
Hypothesis 100
  alternative 100, 101
  null 100, 101
  one-sided alternative 101, 105
  two-sided alternative 101
Hypothesis tests
  χ^2 (Chi-squared)
  for goodness-of-fit 167, 197
  of independence 131, 194
  connection with confidence interval estimation 110
  distribution-free 110
  Mann-Whitney U 116, 123, 125, 194
  non-parametric 116
  sign 117, 118
  Spearman's r_s, 146
  t
  correlation coefficient 142
  difference in means of two populations 109, 192
  mean of population 104, 192
  mean of population of differences 108, 192
  slope of regression line 159
  Wilcoxon signed rank 116, 120, 122, 194
  z 106, 119, 122
Independence, statistical 44
Independent
  events 44
  trials 53
Inference, statistical 77, 100
Intercept of regression line 154
Inter-quartile range
  grouped data 33
  ungrouped data 33
Interval estimation 85

Line chart 17
Linear correlation 143
Linear regression 153, 196
  analysis 153, 196
  confidence interval for predicted value 156, 196
  confidence interval for slope 165
  connection with correlation 160
  equation 153, 154
  hypothesis test for slope 159
  intercept of regression line 154
  slope of regression line 154, 159

Mann-Whitney U test 116, 123, 125, 194
  difference between medians 123
  large sample test 125, 194
  table of U values 246
Mean
  arithmetic 24, 183
  binomial distribution 55
  normal distribution 67
  Poisson distribution 60
  population 86
  relation to median and mode 28
  sample
    grouped data 25
    ungrouped data 25
Measure, numerical 24
Measure of central tendency 24
Measure of skewness 35
Measure of uncertainty 39
Measure of variation 30
Median, sample
  grouped data 26
  ungrouped data 26, 183
MINITAB 181
  list of commands 198
Mode, sample
  grouped data 27
  ungrouped data 27
Multiplication law of probability 43
Multiple regression 153
Mutually exclusive events 46

Negative correlation 142
Negative skewness 29
Non-parametric tests 116, 148, 194
Normal distribution 67
  approximation to binomial 72
  goodness-of-fit test for 173
  mean of 67
  probabilities, using Minitab 188
  standard deviation of 67
  standardized 69
  table of distribution function 240
  table of upper percentage points 241
Null hypothesis 100, 101

Observed frequency 132, 168, 170, 172, 174
Odds and probability 42
Ogive 16
One-sided alternative hypothesis 100, 105
Outcomes of a trial 40
  equally likely 40
Outlier 143

Paired samples 93, 108, 118, 120, 193
Parameter 53, 58, 67
Pearson's correlation coefficient 139, 195
Percentage and probability 42
Pictogram 19
Pie chart 18
Pilot survey, experiment 90, 92
Poisson distribution 58
  approximation to binomial 60
  goodness-of-fit test for 171
  mean of 60
  probabilities, calculation of 58
  probabilities, using Minitab 187
  standard deviation of 60
  table of probabilities 236
  use of table of probabilities 60
Polygon, cumulative frequency 16
Pooled estimate of variance 95, 109, 193
Population
  definition of 77
  mean (μ) 80, 86
  standard deviation (σ) 80
Positive correlation 142
Positive skewness 29
Power (of a test) 100, 110, 116, 139
Predicted value in regression 156
  confidence limits for 156
Probability 39
  a priori definition 40
  addition law 45
  and odds 42
  and percentages 42
  and proportions 42
  conditional 43
  multiplication law 43
  range of possible values 42
  relative frequency definition 41
  subjective 43
  trees 47
Probability distributions 52, 65
  binomial 52
  continuous 65
  discrete 52
  normal 67
  Poisson 57
  rectangular (uniform) 71
  simple proportion 167
  t (Student's) 88, 89, 103, 105
Product moment correlation coefficient 139, 195
Proportion and probability 42

Quartile
  lower 33, 184
  upper 33, 184

Random assignment 93
Random (at random) 41
Random number table 242
Random sampling 78, 187, 189–92
Range 34
  inter-quartile 33
Rank correlation 146
Ranked variable 2
Ranking 26, 120, 123, 146
Ranks, tied 120, 124, 147
Rectangular (uniform) distribution 71
Regression, *see* Linear regression
Relative frequency definition of probability 41

Residuals in regression 158
Residual standard deviation 158

Sample
  definition of 77
Sample inter-quartile range 33
  grouped data 33
  ungrouped data 33
Sample mean 7, 25
  distribution of 81
  grouped data 25
  ungrouped data 25, 183
Sample median 26
  grouped data 26
  ungrouped data 26, 183
Sample mode 27
  grouped data 27
  ungrouped data 27
Sample size 78, 80
  choice of 89, 92
  effect on confidence interval 89
Sample standard deviation, 30
  grouped data 31
  ungrouped data 31, 183
Sampling distribution of the sample mean 81, 189
Sampling method 78
  random 78
  stratified 80
  systematic 79
Scatter diagram 20, 140, 142, 144, 145, 147, 154, 195
Shape, measures of 35
Sign test 116
  for median of a population 117
  for median of a population of differences 118, 194
  large sample test 119, 194
Significant figures 8
Significance level 100, 102
Significance testing, *see* Hypothesis testing
Simple proportion distribution,
  goodness-of-fit test 167, 197
Simulation in Minitab 187, 189–92
Skewness
  measures of 35
  negative 29
  positive 29
Spearman's correlation coefficient 146
Spearman's rank
  calculation of 146
  hypothesis test for 148
  table of values 248
Standard deviation
  binomial distribution 55
  normal distribution 67
  Poisson distribution 60
  population 80
  residual 158
  sample
    grouped data 31
    ungrouped data 31, 183
  sampling distribution of the mean 81, 189
Standard error of mean in Minitab 184
Statistical
  independence 44
  inference 77, 100
Statistics 1
Stem and leaf displays 20
Stratified sampling 80
Student's distribution *see t* distribution
Subjective probability 43
Summarizing data 11, 24, 183
Symmetry 29
Systematic sampling 79

t distribution 88, 89, 103
t table of percentage points 243
t tests, *see* Hypothesis tests
Tables 11–6
  contingency 15
  cumulative frequency 13
  grouped frequency 11
  statistical 233
Tally marks 12
Tests of significance, *see* Hypothesis testing
Test statistic 100, 102, 103
  calculated 100, 102
  tabulated 100, 103
Tied ranks 120, 124, 147
Time-series graph 19
Transformations 161
Trials 40

independent 53
Tree diagrams 47
Trimmed mean in Minitab 184
Two-sided alternative hypothesis 101

U test (Mann-Whitney) 116, 123, 125, 194
Uncertainty, measure of 39
Ungrouped data
  inter-quartile range 31
  mean 25, 183
  median 26, 183
  mode 27
  standard deviation 31, 183
Uniform distribution 71
Unimodal distribution 28
Unpaired samples 94, 109, 123, 125, 193
Upper quartile 33, 184

Variable 1
  categorical 2
  continuous 2
  discrete 2
  quantifiable 2
  ranked 2
Variance
  sample 34
  pooled estimate of 95, 109, 191
  population 95
Variation
  coefficient of 34
  measures of 30

Wilcoxon signed rank test 116, 120, 122
  large sample test 122, 194
  median of a population of differences 120
  table of T values 245

Yates's continuity correction 132, 195

z value 69, 106, 120, 122
  tables (normal distribution) 240, 241
  tests *see* Hypothesis test